Contractor Heaven

Contractor Heaven

Bringing Out the Best in Your Home Improvement Contractor

The Art of Hiring, Managing and Paying Professional Help

Contractor Heaven

Lynn Hartwig

Gettingbestprice.com

Copyright © 2014 by Lynnette Hartwig All rights reserved.
No part of this book may be reproduced, without permission by the Author.
Reviewers may quote short excerpts in a review.

Requests for permission should be addressed to the publisher:
Current Tech Permissions Department, 12018 Quito Rd. Richmond VA 23112-3771
or by email to Lyn8nn@gmail.com.

ISBN: 978-0692273203 (paperback)

Contractor Heaven: Bringing Out the Best in Your Home Improvement Contractor / Hartwig, Lynnette
    p.   cm.
  Includes Index
    1. Dwellings—Maintenance and Repair. 2. Dwellings—remodeling
    3. Construction contracts. 4. Contractors. 5 Consumer Education.
    I. Title   643.7'—dc20   2014914746

Printed in the United States of America

20 19 18 17 16 15 14 13 12 11 10 9 8 7 6 5 4 3

Limit of Liability/Disclaimer of Warranty: While the author has devoted many years to the acquisition of this information, there are no representations or warranties with respect to the accuracy or completeness of the contents of this book and specifically disclaims any implied warranties for infallibility of advice or fitness for a particular purpose. Neither the author nor anyone involved with the publishing and distribution of this book shall be liable for any loss or damage arising from information in this book.
Examples used in this book are altered to protect privacy.

This publication is designed to provide interesting and authoritative information in regard to the subject matter covered. It is sold with the understanding that the author is not engaged in rendering legal, accounting, financial planning or other professional service. If legal advice or other expert assistance is required, the services of a competent professional person should be sought.

This book is available in digital, e-book and paperback formats. While efforts are made to render the content of all formats alike, the author regrets any variations appearing in the differing media and publishing methods.

This book is dedicated to tradesmen,
honest and skilled as the day is long.
They're out there; I find them every time I look.

Table of Contents

Ground Level	1
Chapter One	5
Before It Hits the Light of Day	23
Contractor Heaven	34
Second Quote Phenomenon	37
Finding the Contractor	40
Professional organizations	45
Handyman or Contractor?	46
The Estimator Visit	51
Jargon	55
Time and Materials	61
The Request for Quote—RFQ	77
Content of the RFQ	79
RFQ Special Situations	85
Ethics	88
Payment schedule	95
Milestones	97
The Holdback	100
Signoff at Completion	100
Contracts	102
Keeper of the Verbals	105
Writing the Contract	107
The Law Loves Handwriting	109
Will, Must, Shall	111
Licensed, Bonded and Insured	112
Lien Release	114
Contract Omissions	116
Watching the Work	118
Building Inspector	127
Getting Your Money Back	130

Not Actionable	132
Don't Go to Court	133
Debit Cards in Real Life	136
Credit Card Court	137
Complaints and Warranty Work	138
Major Renovations & Additions	143
Project Scope & Change Notices	146
Hiring an Architect	149
Add On or Move?	152
Great Home Features for Pennies	156
Schadenfreude	187
Definitions	197
Addendum A –Forms	204
Addendum B –Risk	217
Index	227

Acknowledgements

I would like to thank the businesses, agencies and individuals whose contributions informed this book:

In alphabetical order:
   AIA.org American Institute of Architects
   Angieslist.com
   BBB.org Better Business Bureau
   Boston.com (Boston Globe)
   dllr.state.md.us/license/mhic Maryland Division of Professional Licensing
   generalcontractorlicenseguide.com
   Greenbuildingadvisor.com
   Google.com
   Hgtv.com (Home and Garden TV)
   Homedepot.com
   Local.yahoo.com (local reviews)
   Lowes.com
   Menards.com
   Pbs.org (Public Broadcasting Service)
   Thisoldhouse.com
   Yelp.com
   Youtube.com

Very quickly, cell phones, online research, and innovative new materials and processes have changed the home maintenance landscape. Lucky for us, it is easier than ever to obtain prices which are fair to both customer and contractor, easier to monitor the work and easier to detect complications before work starts.

Unlucky for us, the shrinking of the high wall between 'white collar' and 'blue collar' castes since the 1960s combined with the retirement of the last fellows from the old system makes it harder to find high-caliber blue collar workers, since the bright ones hop the fence to white collar. Hiring competence takes more effort than it used to; it's not merely nostalgia blurring recollection. Today, the good ones want to make white-collar wages, and frankly, they deserve it. In the long run, wage equity is the only way to restore high caliber workmanlike quality to the building trades. Making a good living, however, is not synonymous with making a killing.

Contractor heaven is about the good guys thriving because they have happy customers.

Ground Level

Owning property can be one of the greatest sources of pride in our lives. Our home is evidence we did something right; it stamps us as full-fledged adults. Our house grounds us, figuratively and literally.

American culture related to housing can be epitomized in one word: upgrade. Whenever we run into a little money, we upgrade. If we gain a good chunk of money, we upgrade to a better house entirely. Otherwise we just enhance, add on or replace aged or unpleasing parts of our house in piecemeal fashion over our entire lives.

This is not true everywhere. Between France, Italy, Switzerland, Germany and the UK there are over four million houses whose exterior is virtually identical to their 1850 appearance. In the United States, you would be hard pressed to find even ten houses that have the same footprint and look identical to their 1850 appearance. It's in our DNA to fix the old and install the new.

A 2013 CNN Money Magazine survey showed that the typical homeowner averages $4,000 per year on home improvements. Undoubtedly, a significant percentage of those funds are lost to inept work, opportunism, or unnecessary tasks added by the contractor simply to puff up the price. That's not even counting outright fraud, the kind which should be reported to the authorities so they crack down on those offenders. What authorities? When? Ah, read on.

All home repairs and renovations are a service. You are not buying an object from a store, where shopping for the lowest price makes perfect sense. We can buy a Snickers® bar from the grocery store in a pack of eight for a real bargain, or from the hardware store or gas station checkout line for a bit more, or from a vending machine for even more. The taste and calorie count will be identical. The only difference will be the price we pay.

Throw that thinking out the window when engaging a contractor. When you hire a contractor, you are buying the hands-on work of an individual, one who cannot provide more than his own skill level with the equipment he knows how to use—no matter what you pay him. There is no Snickers bar in the construction world. Every single contractor will give you a different set of skills, a different work pace, a different attention to detail, and a different level of cleanliness at the end.

To get the Snickers bar that you want, you need to find a contractor who makes Snickers bars all the time so he can tell you what goes into making one and can show you a couple of pretty good Snickers bars he's made recently. You can take a look at his past Snickers bars, count the nuts, check the thickness of the caramel, make sure the chocolate was applied evenly the whole way around, and see if the material used in the wrapper does the job and looks right.

In short, you want to hire experience. Someone who has done this very thing several times before. The M&M's® guy or even the Payday® bar guy is not going to be right for you.

This applies whether you're hiring a handyman, having a dead tree removed, getting your kitchen remodeled, or having an architect design a whole home. Your satisfaction with the work habits, neatness, skill, error level, punctuality, duration, and price depend upon how well the boots-on-ground workers do your particular job. The Supervisor of THIS crew, the strength, dexterity and experience of THIS worker, these are all that matter. Company reputation or good references mean nothing compared to this. A good reference is relevant only if you get the same supervisor and the same workers; anything less and all bets are off.

How does a homeowner get good work for a fair price? It's easy, yet hard.

Hard, because the contractor[1] negotiates price several times a week, so has done a bit of trial and error on what to say. He has discovered phrases, arguments and tone of voice that work for him. You are not so polished and not so confident. It's not easy to tell if what he says is true or just well-rehearsed.

Hard, because most of what's written on the internet is written by the contractors themselves. I seldom read an internet article by a contractor which doesn't contain a couple of self-serving bits peppered in with otherwise rock-solid advice. I am not a contractor, so I have no horse in the race. That means you will likely find at least twenty truths in this book that you have never heard before, starting with quick ways to spot the wrong contractor, how contractors set prices, the best way to keep your job progressing, building materials to avoid, and ending with no-cost ways to resolve issues without taking him to court.

Often, books and articles purporting to save money on home repairs and upgrades simply advise you to Do It Yourself! What a letdown for most of us, who do not have the tools, hand strength, talent or time to be tearing

---

[1] The terms vendor, contractor, estimator, GC, General Contractor, shop, firm, and business are used interchangeably in this book

out walls, laying tile, or any of it, really. Besides, DIY is first-timer work, and first-timer work has lousy quality. In my house, where I will look at it every day, it must be a great job, not a first-timer job.

Often DIY looks cost-effective and easy only when you know very little about the process. Things that are perfectly safe with a certain amount of hand strength and experience are guaranteed to go afoul without them.

Even if you watch a video, your job will not proceed like that. A year ago I watched a Youtube video on a 'simple' clock repair I needed to do. It took him three minutes but took me an hour. And I had the right tools! But first I missed a step so had to undo it, and then I didn't align it properly, so had to undo it. Worse, I dropped a tiny screw into the clock, so there went ten minutes, poking it with a bent coat hanger wire until my telescoping magnet could grab it. Never underestimate the ability of an experienced person to make it look so-o-o easy.

Many advocates for DIY projects greater than installing a shelf or changing a wall switch do it for a living; another large batch are merely full of themselves after one or two bigger projects turned out 'OK'—often by random luck, as their story reveals. Reliable statistics on DIY projects don't exist because the people who ended up with an unsatisfactory eyesore don't blab about it. DIY projects that look bad, need to be completed by a professional, or actually reduce home value are the norm. The people who do well with DIY are those who have years of hand tool and on-the-job measuring experience. It's a sick joke for journeymen and people with years of hand tool experience to tell untrained people 'anybody can do it.' When yours turns out badly, you blame yourself—and you shouldn't, because they fibbed. Yes, I do DIY, but I'm a Journeyman Toolmaker and was a Habitat for Humanity crew lead on rehabs for a few years. I hammered, measured, cut, and fastened for a living, and then for fun on the weekends.

I am pretty handy with tools, but I think very hard and look things up before deciding to do it myself. Once I even hired a painter to paint two rooms. Why? Because there was a cathedral ceiling. Just buying or renting the platforms, extenders and brushes would put a big dent into the cost savings. Worse than that, I'd be working all by myself eight feet off the ground, reaching overhead, losing my balance once an hour and sometimes dropping things. I knew enough about it to know I wasn't buying a paint job when I hired a painter, I was buying no-trip-to-the-emergency-room.

But above I said that it's easy to get good work for a fair price. I suppose you want to hear more about that. It's easy, because all you have to do is 1) hire experience; 2) not overpay. Both of these are accomplished with a spurt of effort before you call an estimator and a couple of sentences during the estimator visit. That's 50% of the total effort needed. The other 50% is spent fixing up the contract and watching the work, which includes reiterating important contract points at the start of work and maintaining a presence during the job.

Skimping on the before-calling work, i.e., 'the setup,' makes the odds of being happy with the schedule and outcome pretty dinky. A California consumer agency collected statistics from homeowners with projects costing over $50,000, and found that only 20% were happy with the outcome. Over 50% were *extremely* unhappy. A significant percentage lost huge amounts of money to false starts. They initially hired a contractor who took a down payment and then absconded with it before doing much work, or they had to fire their contractor in mid-job.

I'm going to tell you how to land in the happy 20%. It doesn't take crossing fingers and hoping for the best; it takes setup, meaning a bit of research and being prepared for the process before you call your first estimator. The good news is more than 20% can land in the happy. Some of those contractors with unhappy customers were simply doing unfamiliar tasks; when a contractor sticks to his or her bread-and-butter work[2], it's possible to end up with 90% happy customers.

The lion's share of your efforts should go into the initial vendor selection and getting the contract right. There is no time savings by skimping on setup. It's more effort, hassle, and stress when you push the learning into the middle of a job instead of before the start. There is little chance of a happy outcome if you hire an inept contractor. There is no undo button when you realize you forgot to mention something in the contract. Every correction done in mid-job will cost huge amounts of time and money compared to the little bit of low-stress work it would have taken to add it to the scope at the beginning.

The next chapter contains two contractor-hiring stories that start out very normally, perhaps exactly as you might proceed. One involves a minor home improvement, adding a patio door, and the other a medium home improvement, a kitchen remodel. As these projects go off the rails, you may feel the turn of events were totally unpredictable. After each story what really happened and what the homeowners could have done to prevent it is explained. But if the point to prevent it is in the rear view mirror, as sometimes happens to us all, then approaches for rescuing the project are described, enabling them to get something satisfactory in the end.

The rest of the book gives you the tools to use for both preventing and handling issues that arise. This entails redefining what a 'smooth project' is so you don't expect something there is no hope of getting. My projects tend to go smoothly, but don't imagine for a moment it happens because I throw around trust and money like candy at a parade.

---

[2] Explanations for unfamiliar words and jargon words are in the Definitions section in the back of the book.

Chapter One

Patio Door

One day in March, Mary decided to improve her 1960's ranch home. An intelligent and educated woman, she was not an amateur at negotiating. She had superior business skills gained while earning her MBA, win-win training gleaned from expensive consultants at work, and lessons learned from several best-selling books on negotiating. She was a pro at negotiating; a small contractor didn't stand a chance of giving her anything but a rock bottom price.

Mary had a clearly defined project. She wanted a patio door installed in the dining room where a pair of windows were now located.

Mary had a big advantage: her brother-in-law Sam highly recommended a contractor, Simone Exteriors, Ltd. Sam knew a lot about fixing things. Before he hired them last year, he researched similar projects online, got several quotes, checked licenses with the state and ratings on BBB.com. He stated more than once that they were 'the best' for both price and work ethic. He was very happy with his new siding and the two new windows they installed.

Mary phoned Simone Exteriors to set up an estimator visit. The estimator, Hans, was friendly and polite, arriving with a fat briefcase full of brochures of siding and window trim choices. Mary took him to the dining room and told him about the project she envisioned. Making small talk, Hans asked her how she picked his firm. Mary told him about Sam's glowing recommendation.

He had no patio door brochures to show her, but said he would provide some tomorrow. He stood and listened, lips pressed together, for most of the visit.

Towards the end of the visit he got talkative. He said the installation required cutting the aluminum siding, with a possibility of distorting it around the edges. He recommended replacing the siding on just that one side of the house, to ensure a fine finished look. He explained that although it would be newer, it didn't need to match the other sides perfectly. No one sees two sides of a house concurrently in the same lighting. He made an articulate case for that being the best way to handle it. To set any concerns to rest, he gave her the addresses of two properties nearby where his firm replaced only one or two sides, not the whole house, so she could see for herself how well that worked. He said due to the extra care and fixturing required during cuts, she wouldn't save much money if she declined the new siding.

On reflection later, Mary realized he spoke mostly about the siding and nothing about the patio door. At the time though, she was very impressed with his knowledge, his concern for quality, and his agreeableness. He took measurements, wrote notes, and said he would get an estimate back to her the day after she picked a patio door. The next day while she was at work he hung several brochures showing patio doors in a bag on her doorknob. She looked them over, and two days later phoned him with her selection. She wanted the one on the top of the second page of the blue brochure.

She waited for his pricing so she could initiate her negotiating. He emailed a brief proposal/contract with a total price of $7,700. That seemed high considering the patio door set she picked out was $1600 retail, but then she was getting one fourth of the house re-sided by 'the best.'

Now Mary employed her negotiating skills. She worked the bid down to $6,800 in exchange for agreeing to pay 50% upon signing. A big win simply for paying a bit more a week early, Mary thought gleefully. After the price was settled, Mary discussed the timeframe. She wanted the work to start when it was warmer outside. Her wish was an early May start, but Hans said he could hold this low price only if work started next week, "...because material prices go up in May and we have to beat the price increases." She successfully negotiated a compromise April 22nd start date. The contract they presented was brief. The patio door was described by manufacturer, color, handle type and size, but no model number. The other four sentences were about the siding and clean-up. Below that it said 50% due on signing, plus an unexpected 20% due upon the first day of work, and remainder on completion.

The 20% surprised but didn't faze her; since it was only a two or three day project, it didn't make much difference to her finances. On the very bottom a line said "Cash or check preferred. $2,000 maximum permitted on a credit card." Not wanting to bother with two payment types, Mary paid $3,400 with a check.

On April 3rd a man who introduced himself as the President of Simone Exteriors phoned to say the materials have come in; he has another big project pending so could he start three days from now? He would give her $100 off for the earlier start. Because the temps were still dropping to mid-30° F. at night she hated doing that, but begrudgingly agreed.

On April 6th a three-man crew arrived at 8 AM and proceeded to strip the siding off the back of the house. At 9 AM a truck dropped off a dumpster in the driveway. When all the siding was off, one of the crew cut a patio-door-size hole in the wall around the windows. All the removed siding plus wall section, still containing the old windows, were hauled to the dumpster. At noon the workmen staple-gunned a big tarp snugly over the hole, left for lunch and didn't return. She called the office around 1:30 PM to ask if they were coming back soon; her living room and kitchen were very cold. The woman at the other end assured her they were coming back. At 4 PM Mary called again. This time she was told they would be there first thing tomorrow.

The next morning the three workmen were back with the boxed patio door. Measuring, sawing, and hammering went on sporadically for three hours. They seemed to be framing, then removing framing by dribs and drabs. They studied several loose sheets and discussed them in whispers for the better part of an hour without making much progress.

Doing as she was told, she stayed out of the work area for safety reasons. She glanced momentarily as she got coffee refills from the kitchen and peeked out the bedroom window three times. Around lunchtime, after a long period of quiet, she stepped out to find they had pushed the patio door into the hole and tacked it in place. The tarp was stapled up over it and they were gone. How long had they been gone?

Again they didn't return that day.

The patio door didn't seem to sit right. The door sill was higher off the floor on the handle side than the fixed side. The drywall was torn in a ragged fashion up to a foot away from the two sides of the new door. Strips of daylight peeked through between wall and frame in a few spots. But the worst thing of all was the patio door itself. It didn't look right.

She fished out the brochure again and spent a long time looking at the thumbnail photo on the second page of the blue brochure and the one in her wall. The frame looked slightly different. Plus the one she picked said 'screen door,' in the description but this one didn't have any. Maybe it will be installed last. Of course, that makes sense.

She phoned Hans to mention her concerns. When she said it appeared to be the wrong door, he reassured her it was the one she requested. The manufacturer often changes small things from year to year, he said, but they continue to use the old brochures until they're used up. "It means you have a newer version than the one in the brochure."

Mary asked, "Why doesn't it fit well? The door sill is raised on one side. What are they going to do to make it sit flat?" Hans explained "In old houses it is never a perfect fit like on a new house. A good carpenter can fix that with a modified jam." She asked, "Can't they just reset it so it's level?" He replied, "Your joists have bowed over time. We can tear out the flooring to replace the joists if you want it to line up perfectly." He described a major tearout and days of work to simply align the bottom edge to the floor. "We can do that for about, oh, $8,000, if you really want to. But a good carpenter can compensate with trim and nobody will notice." Feeling guilty that she was coming off as a worrywart and micromanager, she apologized for bothering him. Of course they had a plan; lucky for her she was dealing with good carpenters.

The workmen didn't return the next day. Or the day after that. She wondered if she was being punished for calling the office to ask where the workers were and questioning their skill. The weekend came. And went. Would calling help, or just cause the punishment to go on? Her house still needed siding.

On Wednesday morning she called again. "They should be there today," she was told. No one appeared. She called again at 2 PM. The woman on the phone said "They got held up on another job so they'll be there first thing in the morning." That's all she would say; everybody was out so no one could speak with her now. It stretched to two weeks. With no siding on the backside of the house and the patio door not fully installed, it felt unsafe. Her stomach was in a knot every time it rained, thinking it might bleed through the drywall. It dawned on her that she'd already paid 70% of the full cost, had no siding, the patio door was just propped in place, her dining room wall was a disaster, and if they never came back, every recourse open to her had her living like this for months or spending way more money. Mary cried.

The more she looked at the patio door, the more she thought it wasn't the same door she selected. She fished out the brochures to look again. She found this door on the second page of the turquoise "Supreme" brochure and not the blue "Premium" brochure. There was a $600 difference in the price of the doors! Was she ripped off, or was the mistake her own? She had no way to tell. Both were the same size from the same manufacturer. Both fit the description in the contract. She felt sick about it. There was nothing she could do. She was stuck with the cheaper patio door.

On April 20th, Mary phoned her brother-in-law to intercede on her behalf. "Geez, patio doors are not windows." Sam said. "and besides, they only swapped out a same-size window on my house, just slid it in. They are siding guys, they probably never installed a patio door from scratch in their life." But to be helpful he phoned them. The receptionist took a message, but no one returned his call.

On Saturday, April 24th, Mary walked to a neighbor's party. All the houses on this street were built by the same builder, with the same siding and without patio doors. Sipping her wine, Mary found herself admiring their patio door, which appeared to be much higher quality than hers. The hostess said they put it in last year, replacing a single door in that spot. She said "We paid a lot, almost $3,500, but we wanted the best door they had with built-in locks and a sturdy screen door. Can you believe two guys installed it in one day?" Mary stared hard at her drink for a minute. Then she excused herself so she could go home to cry some more.

Monday, the crew returned to install the siding. They hammered and banged for four hours, and at 1 PM the leadman handed over a work completion form for her to sign plus the bill for the remainder. The bill did not reflect the $100 discount for the earlier start date.

Mary was flabbergasted. She blurted, "The job isn't done. Look at my dining room wall! It's a mess. The wall is wrecked. And the sliding screen door still hasn't been installed The sill is still raised on one side and Hans said a good carpenter would fix it. Why are you asking me to sign that the job is done?"

The leadman replied "We don't do drywall or interior trim. You have to hire somebody else for that, or do it yourself. A screen door wasn't included in this contract. Sometimes people don't want those and you didn't ask for it. See," he said as he pulled a copy of the contract from a folder, "those things aren't included. Our contract is for exterior siding and a patio door."

Mary said, "What about the $100 price break for the early start?"

The leadman looked puzzled and replied he knew nothing about that. The invoice is generated in the office, and to his knowledge jobs start when the materials come in, after giving the homeowner at least 48 hours' notice and securing approval. He looked over the contract and asked her where the start date is mentioned. It wasn't in there. He offered to phone Hans to find out if a discount was missed.

Mary replied that it wasn't Hans, it was the President of the company who called. The leadman looked at her quizzically. "That never happens, he never deals with scheduling."

How did her excellent negotiating skills lead to such an awful outcome? Could a more win-win approach have helped? What if she bargained harder to get them down to $6,000? Or even $5,000?

Would negotiating with someone higher up in the company, not the estimator, have worked out better? Did this turn out badly because she didn't wear her power negotiating clothes? Does she need more training in how to appear confident? Could more eye contact and better powers of persuasion have made the deal go smoothly?

You already know the answer. Negotiation skills are silly when applied to contractor situations.

What Mary wanted was a high quality patio door installed in one day for a fair price. What she got was: a low quality door that was installed incorrectly, water damage to her back wall that will cause mold issues within five years, living in fear for two weeks and overcharged to boot.

What did Mary do wrong? What should she have done?

First: She picked the wrong vendor. This is 90% of the problem. She took a recommendation from a trusted person on something she knows nothing about. She couldn't discern that what appeared to be minor differences in the task were actually huge differences. It is fine to take references, but they must be for the exact same work.

Mary had the ability to easily find references for exactly the same work. Everyone who lives on a street where the houses were built concurrently has that same wonderful ability. Simply walk your street and ask the neighbors. People are surprisingly forthcoming on whether they were happy with their patio door installer, roofer, bathroom remodeler or the guy who built the sunroom. Who not to hire is valuable information also.

Two: Payments got ahead of work. Your boss pays you after you put in the week. As the employer of this contractor, pay for every hour and every material only AFTER it is consumed on your job. Paying too early means it's free money if they can make you fire them.[3] This may have been what Simone Exteriors was trying to do with the long delays. Mary revealed she was worried and scared, so they decided to see if she would get someone else to complete the job. Then Mary's only recourse would be small claims court—which would take months, and they would present paperwork showing they incurred costs so owe her little. Because they are a reputable business, they waited her out only two weeks. A less-ethical one would have given it months.

Three: Her expectations were not mentioned in the contract. She assumed a screen door, drywall and interior trim were part of the project even without any mention of those things in the contract. In her mind she was paying for a 'whole job' patio door, and signed the contract they offered without asking for mention of the missing aspects. Now she cannot sue for something beyond the scope of the signed contract. In this case those things were not even discussed, much less promised.

As for the cheaper door being installed, Mary attributed it to a communication error that was half her fault. I'm not so generous. An estimator who hands over six brochures from two manufacturers knows the customer must state the model and ID number from the brochure. When a customer picks a specific item, that ID number is written in the contract. Anything

---

[3] When a contractor asks for a large upfront payment on a short-duration job, it means he already knows he will not finish for weeks.

deviating from that is pure intent to swap out with a cheaper material. Which is what happened.

Additionally, when she was offered an early-start discount, she should have asked for it in writing—an email would suffice. Contractors know that office workers are overly impressed with titles, so Hans asked a coworker to pose as President on the phone to shake loose a concession. He's probably done it a dozen times.

Four: She got all the information from the seller. She could have Googled 'installing patio doors' or watched a Youtube video to find out how long it takes and if re-siding was really necessary. Even half an hour on Youtube would have made her a smart shopper.

Five: She didn't respect his specialty; she figured if he has the right tools, he's right for the job. Just because he has hammers, saws and tape measures doesn't mean he can do anything. There are a dozen firms out there that install patio doors every week; don't settle for a first timer. Ask for references, and go take a look at those installations to make sure they are what you had in mind.

Six: She let them increase the scope of work. She let them charge for re-siding one quarter of her house because they like to put siding on houses. Again, a little time with Google and another four bids would have revealed a skilled crew can cut the hole perfectly, frame the opening per code, slide that puppy in and fix the drywall within two days. Frequently, if you call the wrong contractor for a job, he will find a way to include some of his bread-and-butter work.

When the scope grows into some other feature nearby, something prominently featured on his business card or brochures, it ought to tickle the hairs on the back of your neck that it could be profitable but unnecessary work. Getting multiple quotes will reveal this ploy; one or two might try it, but the others will tell you it's unnecessary.

Seven: Negotiating skills are good for knocking 10% to 20% off the price. In this case she was being overcharged more than double. The paltry 12% reduction she finagled should have been 50% to 65%. The back-and-forth type of negotiating can't get you that low. When dealing with an estimator, the time to lock in a fair price is BEFORE the price hits the light of day. How to do that is in the next chapter.

The price he offers will not be based on what you can afford. Being poor doesn't lower the price; being rich doesn't raise the price.[4] Put that out

---

[4] For details on what does increase the price, see *Addendum B, Ten Risk Factors* that increase the price.

of your mind. If you have no idea what a good price should be, you'll be overcharged in either case.

Eight: She trusted that a contractor who would be a first timer at this task would turn down the job. First timer work sucks. Contractors aren't doing this for the greater good. They do it for money. They lose no sleep if cold air is coming through the gaps they left or a bad smell emanates from the moldy puddles they sealed into your walls.

Nine: the contract didn't have a 10% holdback in it, and she didn't add it. Overpaying so severely like this may mean a holdback might not make any difference, but it would have given her some clout to insist they send her a drywall-and-trim guy, If they refused, she could use the $680 to engage one herself. That way, at least she'd get the job done for the price she negotiated.

This tale also points out a very common response to customer objections to shoddy workmanship: making up something huge and expensive that would fix it. The easiest way to shut a customer up when they ask for good workmanship is to attribute the cause to some pre-existing condition and then describe expensive and invasive procedures as the 'only way' to correct it. The more destructive the better.

Their comeback is that won't be just three hours of rework consisting of plenty of elbow grease and greater attention to detail, oh no, it requires tearing out beams, foundations, destroying already-installed components or invading up-to-now untouched areas. The homeowner, already getting the feeling these guys are inept, will be frightened to let them tackle a bigger project involving more teardown.

Handling refusal to do rework is what the holdback is for: if they won't realign the crooked part, or fix that gap, or fix the stained part, you tell them to subcontract to someone with those skills or else you'll use the holdback money to hire someone.

Having some knowledge of a few basic terms, a few costs of the main materials plus a rough idea of actual working hours is the way to prevent being charged double or triple the actual value. In Mary's case, using patio door installation terms wouldn't help. The moment Mary revealed to a siding guy that she intended to hire him to install a patio door, he pegged her as a know-nothing. Even the most articulate use of trade terms will not fix that. He didn't have any patio door brochures in his bulging briefcase full of brochures because his company doesn't do that. It was a dead giveaway that Mary missed.

What Mary did was similar to making an appointment to see an Ophthalmologist, and once there, removed her shoe and asked what to do about her bunion. It's easy, but wrong, to think a doctor is a doctor, just use the closest one for everything. Doctors specialize, and so do contractors. Nobody does it all. She revealed she hasn't a clue that Ophthalmologists don't do feet. In the case of the eye doctor, he definitely will say, "I'm not the right medical person for that surgery."

First, Do No Harm—Said No Contractor Ever

Contractors do not take an oath of 'first, do no harm.' They don't have to tell anybody handing them money that they are a first timer. Not when they can pick up a book at the library or read the instructions in the box if they get the job. Which is what happened to Mary. They were so disrespectful of her that they didn't even read the directions before arriving at her house. They did it right in her back yard and quarreled amongst themselves when they didn't understand.

For three grand free and clear, maybe the Ophthalmologist will pop into the library for a book, talk to a friend who's done it before, and take a stab at fixing Mary's bunion; he's already got the scalpels, sterile bandages and antibiotics, after all. Wha?!? It's so unlikely that I'll bet you're laughing. With a contractor, not so funny. He's already got the truck, hammers, nails and saws. If he wants the money, he will tackle your job.

As you might guess, the odds of bunion surgery having a great outcome when done by a first-timer Ophthalmologist instead of an experienced Podiatrist are very small. Ditto the contractor.

The key difference between negotiating with contractors vs. every other kind of negotiating is that the thing you negotiate, the price, is absolutely meaningless if you're talking to the wrong shop. The negotiation isn't a 'win' if you get a good price but don't get a good outcome.

The stress Mary felt during those days when her house was torn apart and she had no leverage at all with the contractor is something you cannot fully appreciate until it's happening to you. Mary had no idea she set herself up for this.

The good news is that you can hire contractors and do just fine on price and workmanship without any personality change, new clothes, quick wit, nerves of steel, hair style change, eye contact, gift-giving, and most of all, no need to figure out a win-win option.

In Mary's case, an hour devoted to watching some Youtube video on patio doors, getting five quotes and adding all her wishes into the contract would have saved her two weeks' of feeling miserable and spending $4,500 more than reasonable for that cheap patio door.

Kitchen Remodel

Sue and her husband Bob decided to remodel their kitchen. Sue asked for my advice in the beginning, but after discussing it, the couple decided they didn't want to go to all that bother of getting multiple bids, googling for average prices on cabinets and countertops, and watching Youtube videos. They agreed they would simply go to the nearby big-box hardware store and let them handle the whole thing.

They both believed a nationwide chain store with a large kitchen department would have experienced workers and would not resort to the shady tricks of small-shop contractors. With a reputation to protect, the national chain would treat customers right. The people who work in the stores are paid an hourly wage, so they have no incentive to jack up prices to make more money. Because they are large, they can provide larger crews all working at once, making it quick. It sounded very reasonable and safe.

Bob and Sue agreed to select only cabinets, countertops, sink, faucet and flooring that were off-the-shelf, high sellers and therefore quickly available. Through much of the selection process they talked themselves out of their preferred styles and choices in materials. Other people paid through the nose for custom touches and stylish looks. Other people had long waits and delivery hassle. Not Bob and Sue; they were going to be smart and avoid that. The kitchen, when done, might not be eye-catching attractive or congruent with their tastes, but it would be quick and stress-free. They had two small children at home so couldn't tolerate the disruption.

It took them a few weeks to agree on the compromises in cabinets, countertops and flooring. Initially, they planned to incorporate some changes in sink and stove placement and a little drywall work that would have greatly improved the layout. When they were told that work would increase the price by $7,000, they decided they couldn't afford it. They didn't argue with the price. They didn't ask, "How the heck can changes using materials costing less than $300 and taking less than 30 man-hours have a labor cost of $6,700? That's over $200 an hour, when $60 an hour makes more sense." Instead, they simply abandoned the much-desired improvements.

An estimator came to their house to measure, and while there, pitched the service and skill level of their installers. During that lengthy visit he described several high quality touches they provide, such as sanding the wall behind the countertop and backsplash to remove the unevenness that usually develops in drywall over time. This unevenness creates unsightly gaps between wall and the hard backsplash, spaces where dirt and oils can lodge, creating bad smells. Sue and Bob were very impressed with that attention to detail. They became even more pleased with their smart choice.

Their cabinets were a standard, non-custom design, but they wanted the top trim to match the wainscoting trim in the dining room. The store had no trim like that, but their salesman found that by laying three standard trim pieces on top of one another it was a close approximation. For 40 feet of cabinetry edge, the big-box store set a price of $2,000 for the trim. When Sue asked why so much, since each of the individual pieces cost under $3 per foot, the salesman said it was due to extra labor. Each piece, he explained, would need to be nailed carefully into place. It would take two workers to align each additional piece while standing on ladders.

They really liked the look, so they agreed to the price. I'm ruining the suspense, but I'll tell you how installing the trim was actually done: the workmen set up a makeshift table using a sheet of plywood. Seven foot lengths of each of the three trim pieces were laid down, glued together with wood glue, and clamped to the plywood. It took about three minutes per set, six sets needed. Including table setup, it took less than an hour. The next day they unclamped, measured, cut, and checked the length by holding the pieces against each cabinet while the cabinets were still on the floor. It took about half an hour to measure and cut all the trim. Later, after the cabinets were mounted, it took minutes for a guy to nail them in place.

Materials: $361.20. Labor: 2 man-hours @ $819.40 per hour.

Back to the story. When they had made all their selections, the project came to $27,000 plus $2,000 for the trim. This was their entire budget. It was uncannily close to the $27,000 budget number they freely shared with the estimator at the beginning. It was a budget number they thought they were moving away from each time they chose a lower-cost material so they could spend more on appliances. By their math, they dropped over $3,000 from a number that initially included appliances. Why had the total grown, not shrunk? Sticking to their plan to reduce hassle by working with only one vendor, they sadly abandoned their intention to buy new appliances. They would make do with their current appliances for another three years while they paid off the loan.

The big-box store offered them a 5% discount for using the store's charge card, so upon signing the contract, Sue and Bob handed over the plastic and the full $27,550 was charged.

During the hardware selection the salespeople described what sounded like a three week process: three days for removal and cleanup, two days for flooring, a few days for cabinet and trim installation, a day for plumber and electrician, a day for countertops, a day for painting, a day for moving appliances back in. Since it was early September, that was acceptable; they were hosting several out-of-town family members at Thanksgiving, so even if it took twice as long, it would be done in plenty of time. They were so confident the big box store was going to be low effort and low stress that they didn't even care that the contract specified no dates and no repercussions for missing those dates.

You have guessed what's coming, right? By Thanksgiving the kitchen had been gutted for almost two months: no cabinets, no kitchen sink, all appliances shoved into the dining room. The flooring was installed, whoopee. Requests for progress were met with 'tomorrow' or 'next week.' Just days before Christmas the main cabinets and countertops were in and the appliances were restored to their positions, but several little things like painting, replacing a broken cabinet door and the backsplashes were not completed until February.

When the installers affixed the long backsplashes, they did not sand the wall. The unevenness was very visible. In some places a stack of eight business cards could slide behind it freely. Bob phoned to request rework. The person on the phone said this is normal in old houses, it's unavoidable. Bob repeated the sales pitch about sanding the wall prior to application. The person on the phone denied that was part of their process.

The person on the phone asked the installer to email some photos of the disputed backsplash. The big box store called the next day with their decision that the backsplash was 'fine.' If the homeowners were still unhappy, they could pay for a complete replacement of the backsplash plus labor.

This was the straw that broke the camel's back; Bob and Sue were outraged. If they had known the sales pitch bore no relation to reality, if they had known the price was going to be their budget number no matter how cheap they went on materials, if they had known nobody gets a three-week kitchen remodel from that store, ever . . . well, then I guess they wouldn't have picked the no bother, stress-free, quick way, would they?

What did Bob and Sue do wrong? What should they have done?

While the elements of what they did wrong are similar to Mary's, the priority is different. The national chain can be a good choice, sometimes. Bob and Sue's belief that the store is immune to both matching the competition's price and profiteering is totally wrong; those have the same power on businesses regardless of size.

Their belief that prices are more likely to be fair because all the employees they deal with are paid an hourly wage, not commission, doesn't hold up to even the briefest scrutiny. Almost every estimator works for an hourly wage, small company or large. Pressure from managers to boost sales 'or else' works just fine. Charging the full budget number, if the customer reveals it, is simply a no-brainer. Meaning, literally requiring no math; why multiply and add up a bunch of numbers when the customer already said what they will pay, and it's way more than enough?

Salesmen may have bonus plans, but these have the opposite effect. If there is a bonus plan, it's usually on sales by count or total dollars, not on just the amount that goes over the base cost. What this means is that salespeople on commission are not inclined to overcharge unless they're 98% certain the customer isn't obtaining another quote elsewhere. Their bonus is higher with two $22,000 sales than one $27,000 sale. If they lose a customer because another firm underbid by $1,000, they get nothing. If they score nothing too often, they get fired.

Companies have many ways to devise an incentive plan, but one thing is always true: when the estimator/seller knows the customer is obtaining other quotes, it is too risky to inflate prices. When the estimator believes there are multiple bidders, he or she keeps the price close to time and materials.

Opportunistically overcharging when there's no competition is called normal business practice. That was the case here. Bob and Sue, by refusing to obtain another quote, encountered normal business practice.

Whether it's a fair price or an opportunistic one, it's a huge mistake to pay in full up front. Even worse is doing it with the only credit card that provides zero clout for non-service: the store's own card. No wonder the store offers a dinky price concession to tempt customers into having their hands tied behind their backs financially. Without even a 10% holdback, customer complaints about quality can be high-handedly waved off; it would take nothing less than a class action suit to get them to refund for skipped installation steps.

Paying by credit card up front might be how the big-box stores work, and they might say their contracts are not customizable by the customer. They can get away with these self-serving policies because they're large. Stores, big or small, are in the business of making money. There's more money to be made by sending barely-trained minimum-wage installers and by showing the revenue on the books for three months before incurring material and labor costs. Depending on a business to resist the lure of financial gain out of goodness is asking too much.

When things start going off the rails with a big-box store, the only tool in your toolbox is to be a twice-a-day pain in the neck[5] until it gets back on the rails. Ask to speak to his manager, and then his manager. Threaten to write reviews, call the newspaper, or hint that you will call that fellow at the TV station who does the consumer segment on Tuesdays. Being a thorn in their side works well because they are paid an hourly rate, not a cut of the

---

[5] The call itself is the burden; you can be pleasant and calm throughout. With caller ID, maybe he doesn't even pick up; mission accomplished. He's not forgetting you; you loom large in his mental landscape.

profits. They don't need this guff. It's no skin off their nose to make you less complainy by giving away the store's money or changing the schedule.

When a big box store doesn't keep its promises, most people resignedly give up, thinking they can't win against such a huge opponent. Actually, the opposite is true. The odds of winning are higher than with a small fry contractor who has already spent the money. If you're willing to make several calls and tell your story plus your desired resolution clearly and relatively unemotionally six times to six different managers at the big box store, the odds are good they will settle to your satisfaction. Don't give up.

Bob and Sue slipped a gear, as do most people, by thinking that a large corporation can match their sales pitches to their actual service in every single store. The brochures and sales training are developed in an entirely different department in a far-away building by people who have no interaction with installers. Sellers are told what to say to customers and have zero responsibility for whether that happens or not.

If the installer doesn't sand the drywall, don't call the guy who told you they would. He can't help you; he never met an installer in his life and has no idea what their training encompasses. He's instructed to tell the story about the sanding, so he does. He's instructed to brag about their depth of experience and how many kitchens they do every year. If the crew that shows up at your house consists entirely of 19-year-olds hired in the past nine weeks, do you think it matters to him? Far from it; he does not even feel like a liar.

Calling two other kitchen remodelers, even if one of them is another big-box store, means playing them off of each other for lower prices and added features. It is not as much work as Bob and Sue think it is. They already did the hard work of selecting what they wanted. Hiking those choices to three other places for a quote is hardly any work at all. We're talking about six items: countertops, cabinets, faucets, knobs, sinks and flooring; there are only a handful of manufacturers in each category and they sell to everybody.

For everything but cabinets, write the manufacturer and model number on a piece of paper. For cabinets, Bob and Sue simply bring a picture or brochure of their chosen cabinet style to the second and third shops and tell them to match it, either by the same manufacturer or something similar in quality and style that their preferred manufacturer makes. They know the count, size and placement of each cabinet type, so it becomes merely a shopping list of parts. Three of these, five of those, one of these, glass front on two of these, and so on. For an extra two hours of their time, one in the store and one with the estimator taking measurements in their house, I'll stick my neck out to say it's impossible the second price is higher than the first.

I'll stick it out even farther to say the second shop will suggest something that improves their enjoyment or meets a need that they were forgetting. Perhaps it's a garbage can location or a pet feeding location; advising them to angle a sharp corner near a door; showing them a counter chair they love.

Maybe the second big box store will say, "If you buy a frig, stove, dishwasher and microwave oven in a single purchase in the next 30 days, you can have 30% off whatever the price is that day, even sale prices." Maybe they can add back in the eight hours of drywall work they lopped off due to opportunistic overcharging.

Being completed by Thanksgiving was important to Bob and Sue. One of the new bidders may be so confident he can clear that worst-case timeframe with their selected materials that he offers to write into the contract: $5,000 off if flooring, cabinets, countertops and appliances not installed/hooked up by Nov. 20th.

Lastly, the really unfortunate mistake was their idea that a huge corporation won't overcharge[6] or mistreat them. I can't imagine where that came from. It didn't come from the movies; it takes about a minute to name fifteen movies about big corporations hurting the little people.

One customer's little backsplash quality issue is simply too dinky to matter. Their policy book is written, and it says no re-doing the backsplash for free. The customer simply has no leg to stand on: it isn't mentioned in the contract. There is company policy saying free replacement isn't allowed. It looks fine from certain angles, isn't cracked, is the size and material that was ordered, and is placed in the proper location. End of story. Unless you are a pain in the neck, in which case what they'll fix is the PR problem you've become and only coincidentally the backsplash.

In this case there's no advice that Sue and Bob should read reviews, check with the BBB and ask for three references. Those are really good ideas nearly all the time. That makes a lot of sense for ten-person firms, even two-hundred-person firms, but not so much for the big-box stores. The big box stores either outsource the installation to small local firms or minimally train installers themselves, so one job tells you nothing about the next job, which might be done by an entirely different crew. Installation errors, not product defects, are the cause of 80% of home improvement problems. If you can meet and talk to your crew from the big box store a week ahead of time, absolutely do that. Without doing that you have no way to ferret out their experience level and if needed, demand a different crew. If the product

---

[6] When necessary, a customer can pay more for speedy work or rush service. If the company charges for that, then you must get it. If they don't finish significantly faster or on the deadline, they cannot keep the money. Any time you pay more for faster service, the contract must clearly state they forfeit the extra if they're even one hour late. I have invoked this at least six times in my career.

fails, incorrect installation probably caused it. When the manufacturer determines installation instructions weren't followed, the warranty is void. Burn!

Experience is the only thing that reduces installation errors. The only way you can hire experience is to hire the boots-on-ground workers who have experience in this exact thing. It's usually better if your estimator will also do the work or be on site to supervise the crew. The second best way is to hire a company that has been doing this for more than five years and has good employee retention, with your lead man having over ten years' experience.

Getting the experienced crew to do your installation is the best way to have the work go smoothly, with few or no callbacks. You might feel uncomfortable asking about all the crews they have and insisting on a certain one. Doing just that, however, pegs you as someone who knows up from down. You'll get more respect and less funny business from your contractor than his average customer does.

From these two stories you have learned a lot already. The main point conveyed by these tales is for home repairs and home improvements, price is not the main thing; job duration and a good outcome are the most important things. A low price is pointless if the job is done badly..

Now I'm going to tell you a little-known fact. A humongous little-known fact. The very core of this book! Here it is: The best way to ensure your job is done right and in a timely fashion is by not overpaying. If you overpay, your job is predisposed to be done badly and take too long.

Did you notice that in both of the above cases, the homeowners overpaid? That was not coincidence; overpaying caused the poor service. Or rather, was a symptom of the underlying cause that would inevitably result in poor service.

I realize that is not intuitive or obvious. Pay more, get more, right? You probably are thinking that for every story of someone overpaying and getting crappy work there are just as many stories of overpaying but being happy with the timeframe and outcome. Not so. I've given you only two stories. There are hundreds of tales about bad projects that start with a too-high pricetag. Statistically, overpaying a contractor does not turn out well.

When is overpaying likely? When:

1) Hiring a first timer who adds in a lot of just-in-case money to cover what he doesn't know. Even if his initial price is competitive and his heart is in the right place, he makes mistakes that cost you money. The completed project looks amateurish.

2) Hiring an opportunist after he found out you're getting only one quote. After you accept a huge pricetag without putting up much fuss, he'll

buy cheap junk and tack on a few cost overruns and whatever else he can dream up.

3) Hiring a newbie who quotes a low price because he underestimates his time or the cost of the tools he needs to buy. He will ask for more money once he starts going into the red. If other work comes his way, he'll go do that first since he needs the money. You'll have to pay him extra or he'll abandon you.

4) Hiring a guy whose business is so in the red is that he's trying to save his company by charging you double or more. He may not pay his suppliers, who will come after you, or he'll simply go out of business before finishing your job. He's just another flavor of opportunist because it only works if he can dissuade you from getting other quotes.

5) Hiring a swindler. At first he's the lowest price, but he demands a high upfront percentage in cash. If he shows up at all, it's to complete the teardown so he can intimidate you into giving him more money, or will do shoddy work that you have to hire another contractor to fix. Don't threaten this guy; just scoot into the bathroom and call 911 while he's still on the premises. Don't worry about being mistaken. The only mistake is waiting until after he robs your peace of mind and takes off with your money. Theft, under the pretense of being a contractor, is robbery.

A generalization about good contractors is that the business was built by people who like the work, not the selling. If the original entrepreneur is still doing the estimating, that sell-dreading inclination is what you have to deal with. He's going to have a style, to put it kindly, that needs some adapting to. If he isn't still doing the estimating, he's hired people to do it, or family members to do it, or a supervisor does it. Usually you aren't dealing with salesmen, you're dealing with someone who wears another hat within the company and does estimating too.

While getting good work goes hand-in-hand with getting a fair price, mere low prices do not lead to good work, because the lowest bid is where the swindler lives. Understanding the difference between opportunists (normal businesses) and swindlers (slimy thieves) is crucial to getting a good outcome.

The final, largest factor you have to deal with is that you don't do this often. Statistically, nearly everyone buying a new roof is purchasing one for the first time. Since a house needs a new roof every 25-30 years, the percentage of homeowners doing it twice in ten years is very small. Even if a particular homeowner is buying their second roof job in eight years, it's for a different house in a different county or state so the pool of contractors and the complexity of work are different. For an addition or kitchen remodel, once in a lifetime is even more likely.

That makes the estimator's job pretty easy, right? We're all newbies and he does this every day. For most of us, a contractor will be in the running for the job if he doesn't raise suspicions, offend or annoy us and has a price in the ballpark of what we expect.

What is your big-picture strategy for hiring contractors?
1) Have a price in mind that's on the low side of time and materials before talking to the preferred contractor(s)
2) Understand what is normal behavior and dress so you don't rule out any good candidates
3) Fix up their inadequate contract to encompass your wishes and expectations
4) Behave like you are the boss. You've hired these employees temporarily and respect their talents and dignity, but you have every right to supervise the work and maintain the standards you prefer (without losing your temper). You don't make suggestions, you express your preference.

In day-to-day life you will meet people who are really attached to their global one-size-fits-all theories about contractors. You've heard them talk about contractors: They're all cheats. They all rip you off. There are no good ones.

I find good contractors every time I look. In fact, I find three to five of them, because I don't stop until that's how many good bids I have in my hand. These aren't perfect, never-make-mistakes, never-try-to-overcharge-me contractors. They aren't always clean-shaven, English-fluent, non-beer-smelling contractors who own all the equipment they need and will not be renting or subcontracting anything. That's too much to ask. I find good contractors, not multi-talented contractors who have taken vows of poverty and sobriety.

To be good, all contractors need oversight; they need a homeowner who clearly states their expectations, and need a contract that includes the estimator's verbal promises plus all the requested work and materials. They need to be paid after they perform the work and not a dime before they've earned it. They need to feel certain this homeowner has it in her to be a PR problem should she be mistreated. Lastly, they need to pull the proper permits and have a tolerable relationship with the town's Building Inspector.

If you deny him any of these needed things, be prepared to have it go sour.

Before It Hits the Light of Day

First, let's talk about not being overcharged.

This is your goal. This works far better than a goal of the absolute lowest price, which can lead to some dark places.

Nancy Lopez's biography contained an intriguing insight on meeting goals. Nancy is arguably the best women golfer ever. She had an approach to winning that is huge for every golfer, and also for every person hiring a contractor. When she golfs, she does not aim for the hole. She aims for an eight foot radius around the hole, a sixteen foot wide circle. Her short game, meaning her putting, is good enough to get it in the hole with one stroke if she can place it within that circle.

Suddenly her game is a whole lot easier. Not so daunting, to drop a golf ball anywhere in a sixteen foot circle. She's figuring out the bounce and dribble to get it into a living-room-sized space, not limiting herself to a rolling trajectory 1" wide aimed at the hole. The latter leads mainly to awful failures: rolling into sand traps, into the rough, you name it. The former is pretty doable nearly all of the time. For a good golfer, I mean.

Getting a fair price from a good contractor is just like Nancy's sixteen foot circle. Conceivably it could be more on the sweet spot. But if you get it within the eight foot radius, meaning you are not grossly overcharged, good game! Often it's within two or three feet, and it might even go right into the hole. Maybe even more often than if you always aimed for the hole. The sixteen foot circle is a good balance between risk and reward.

To not be overcharged, you need to do and say things that will lead your estimator to conclude either his price must land in the sixteen foot circle or he will never win. If you set it up correctly, plop. The price lands in the sixteen foot circle in one stroke. Sure, you might try to work it a little closer to the hole after it stops rolling, if you like.

Your setup work is determining the distance and obstacles between you and the green, plus which way the wind is blowing. If you decide to swing at it without doing any setup work, meaning just call the guy without any idea of the price range for the features, footage and options you want, the odds of getting close are . . . I can't guess. Think it saves time? You'll need to work nine Saturdays for the overtime money to cover the price difference between my way and the no-setup way. Is 72 hours at work a good trade for watching two hours of Youtube videos? Only if your job is playing golf.

Once the contractor's too-high price hits the light of day, haggling or bargaining is good for bringing it down at most 30%. Most will not budge more than 10-15%. Getting only a 30% bump on a job that's 70% overpriced may sound bad, but what's worse is asking for a huge price reduction on the bid from an honest, fair, top notch guy. They are out there.

These are the guys who simply want to make a nice living doing work that gives them pride and joy, without engaging in all the stress and personality-reading that setting prices by individual homeowner gullibility requires. They would rather work steady than overcharge and work less. They do the math and everyone gets the fair price with only a small bit of risk and contingency money added in. These are often the best guys. They are working from the inside out; it's a calling, not a job. I've met cabinetmakers, patio installers, bathroom remodelers and fireplace installers who were like this. They wouldn't allow a homeowner to pick something that wouldn't look good with their décor or that they would regret later. They considered their work to be art. I did too.

Sounds great, right? Except he will never work for you. Not if you don't know what the fair price is and simply approach every contractor situation with a routine counteroffer for 30% off.

He will not counter. He will say no. Result: the haggling approach results in deals with only the grossly-overcharging contractors. You would be asking the ones who presented a fair initial price to work for $5 an hour. A guy with an ethical approach to his work will be offended, so this can't be fixed. A negotiating approach that alienates the most honest and talented contractors does not do you any favors.

The objective is to handle the estimating process so the good contractors don't indulge in opportunism, the scammers weed themselves out, and MOST IMPORTANTLY, you recognize the fair price of the ethical contractors instantly.

When hiring contractors, the important thing is that the good guy gets the job, not the unethical character. This means the good guy thrives. I like the idea of the good guy thriving.

Businesses are in business solely to make money. There are people who quibble with this, citing marketing strategies and advertising blurbs as examples of higher purpose. I snort at this. Let me put it another way: Businesses that do not make enough money go out of business. Any disagreements?

Money is number two only when there is plenty of it. Indulging in 'green' or 'no child labor' or 'free-range' is done as long as it fuels sales.

When it stops doing that, the business stops using more expensive raw materials and procedures, or fails.

No one has a crystal ball on the future. A business cannot be certain a huge expense won't arise four months from now that eats the annual profits and then some. Over-throwing the mark, aiming for a profit point past the target, is the way to ensure there are profits at the end of the year and not red ink.

Therefore, charging what the customer will bear is a greater determinant of end price on a single job than time and materials. It's true of Microsoft and it's true of Max the Chimney Sweep. That's why making time and materials the maximum you'll bear is so successful.

Most contractors, at least the ones we're going to focus on hiring, want their business to grow and thrive for several more years. They want happy customers and good word of mouth. They want no hassle with the law, the BBB, or their professional organizations. This means that 40% to 80% of their customers get a fair price. Only customers who seem an easy mark get singled out for overcharging.

When hunting for good contractors, you don't need to find the ones who never overcharge; they don't exist.

Don't rule out a business who for a minute or two considers overcharging you. They all intend to pad the quote of customers who appear to be an easy mark. Our job is to move out of the easy-mark category *as early as possible* and into the group who knows what the job should cost.

Opportunism is overwhelmingly typical when you reveal you've already decided to give them the job before hearing their price, like Bob and Sue did, or gush about a recommendation from a respected person, like Mary did.

While opportunism is common, it's not a growth-oriented practice. There is little real-world evidence that abusively opportunistic shops make a killing ripping off customers. Mostly they go bankrupt. Going to the dark side is usually motivated by red ink on the bottom line. Businesses making healthy annual profits don't do it on the backs of a handful of opportunistic sales. They do it by being in demand because they give good value for the buck.

In the next five years, Bob and Sue will tell their story and turn their friends and relatives away from repeating their experience. Mary's brother-in-law Sam will never again say a good word about Simone Exteriors. It adds up. Business dwindles. Roughly, for every opportunistic sale today, five future customers are lost. Sometimes one of those jobs was in the very near future, as close as next month.

Even though businesses shoot themselves in the foot by being opportunistic, you can work this to your advantage. Customer goodwill and future business are very important to opportunists since they are likely to be desperate for customers. Throwing future business their way or giving them a lead, such as "my next door neighbor is going to need a new roof soon, too" or "if this job turns out well I'm going to write a review on Yelp.com" is a strong incentive for them to give you a reasonable price as well as work hard to give you a satisfactory project.

To prevent opportunism:

Nip the thought in the bud. The longer he salivates over his big killing on this job, the harder it is to derail that train of thought. After just ten minutes, it may be impossible no matter what you say or do. You've got to say something in the first 90 seconds to knock the notion out of his head. Like what?

One, let him know he has competition. You don't have to be heavy-handed about it. An easy way is to schedule more than one contractor on the same day and let the first one know he has to tie it up by such-and-such time. If two competitors see each other, so what. They already know the other firms are out there.

Two, send a written RFQ. They understand that a written RFQ, faxed, emailed or snail-mailed is going to more than one place.

Three, ask time and materials questions. I'll write more about these later.

Four, if you have done a project like this on a previous property or know someone knowledgeable about this kind of work, let him know very early in the conversation. Wind the conversation around to mentioning that other house, or actual connections you have. For instance, saying that a close neighbor used to build houses in the summer during college or your out-of-state brother is an electrician gives the estimator the message that you already know what a fair time and materials price is, or will find out shortly after receiving his quote.

Five, mention something the last estimator said. A feature or benefit you want to make sure he includes in his quote too is the perfect way to do this.

Six, if you are really, really sure about your preliminary time and materials numbers, go ahead and feed him a number 3% below that sometime before the end of the visit.

Strangely enough, just holding this price in your mind will pull his offer towards it. There have been research studies on this. Scientists speculate it may be due to eye dilation or micro-expressions. I've had real-life experience with the pull of mere brainwaves on the vendor's price, but I don't like to depend upon my aura handling the money part. I'll find a way to work it into the conversation: "On this project, $9,000 is the price to beat." "It's a nice little $2,000 project. I can be very flexible on date so you can work it in whenever, any time in the next 60 days." "I want the $900 one, that's retail, and it will take one guy less than a day, right? So I'm looking for a price like that."

If you've done your time and materials but aren't confident enough about it to say it aloud, visualize your estimated price for a few seconds several times during the visit and let your aura do the work. Strangely enough, often it will yank that price into the vicinity.

Do all six if you can. Anyone who provides a high price after that is highballing, not being opportunistic. Highballing is explained on page 32.

Talking the Opportunist into a Lower Price

What if the too-large cost estimate hits the light of day before you could stop it? This happens more often with women and retirees, but it could happen to anyone. The estimator operates on pre-conceived notions or merely wishful thinking and blurts out a price before the customer can engage their countermeasures.

Sometimes I've been asked "If a company provides a quote that's so high you know he's overcharging, shouldn't you just walk away, he's a rip-off and not worth hiring?" Well, this makes a lot of sense. And I'd probably agree . . . if I wasn't a woman.

Because I am a woman I can take my car to eighteen auto shops and get overcharged at every single one. Yet some of them do good work. Because I am a woman if I ask for pricing on getting twenty pieces of sheet metal bent on a press brake I might be quoted double what a guy would hear. Yet this shop could have 100% on-time delivery and high quality.

No matter what your level of expertise or what you do for a living, the next situation is brand new. This person will be using his pre-conceived notions and wishful thinking to decide whether to overcharge. Seldom does a reputation protect you, even if you're famous.

Do you know who gets into a lot of bar fights? World class boxers. You and I know they didn't start that fight. That's their work. They just put in an eight hour day doing that. Fighting in a bar is like working overtime for free. If there are people out there who want to get punchy with a professional boxer, there are people who will try to overcharge me even when they know I wrote this book.

So put aside the notion that someday you will be so-o-o good at this that that no one will even think of overcharging you. Every project is a new bar, with a new crowd, and there will be a guy who thinks he has you pegged as not so hot.

Occasionally you will barely open the door, say a greeting, and out pops a ridiculous price. It's most likely to happen with external work, where they had a look-around before ringing the doorbell, but sometimes it's internal work, sight unseen. He's testing the waters to catch your reaction.

It's still possible to get a good bid from him, even though he just put his foot in his mouth. The estimator isn't the whole company. They could have some pretty good work crews. This particular company could do great work and be the best one for this job. Can't tell yet. So give him a mulligan.

Use a bit of humor to dismiss that blah-blah he just said. Give him a way out, like "Oops, I think when you look at this job a little closer you'll see it doesn't include that $3,000 of other work most places usually have." Or, "What? I didn't catch that price, but I'm such a visual person, I never remember anything I hear, so please let's not talk price, send it to me tomorrow in an email."

"Oops, perhaps when we get into it you'll see the window can be replaced in less than four hours, with a $300 window."

In one case I said, "Oh-ho-ho, whew, I'm giving you a mulligan on that because I know I don't look it, but I'm a cost estimator just like you. For this project, $X is the price to beat." He laughed, we talked about my career a bit, we did the walkaround, and later he emailed a price about $200 higher than my price to beat. Because we both know the only price to throw out is the rock-bottom time and materials price. The sixteen foot circle.

Do whatever feels natural to you to treat it lightly, pretend you misunderstood, or give him an out.

What if he sends a written quote you determine is out of bounds? Call to say it must be a typo, that part of someone else's quote was on the page or he hadn't modified it fully. All estimators use past quotes for templates and simply change some of the details. It is so common to miss deleting or editing something that it could be the truth.

Fat-fingering an extra zero happens, and while rare, never dismiss a quote from an honest and long-established business due to a hapless typo. Ask him to try again, adding that the other quotes were around $X,XXX less, and you really hope he can be competitive since you were very impressed with him, if that's the case. If it's a bid you were not going to consider anyway, for any reason, then simply set it aside.

You can still have a reasonable business relationship with firms that make quote errors and typos. Typing/writing skills and manual tool skills are two entirely different things. A customer like me who is on the ball and shapes up the quote to meet my own standards sets a wonderful tone for the ensuing work project.

My co-workers say, "Lynnette, you seem to attract more than your share of quote typos." Or maybe other people look at the typos, blanch, and discard it without ever clearing it up. Or maybe they were never typos at all but were intentional overcharges and feature omissions until I called the office to get them fixed.

"Hi Leah! Say I just got the quote, and that line about the contractor shall obtain all necessary permits is missing; the estimator and I talked about that and he said you always do it. We agreed the middle payment would be after passing BI inspection and not the 50% done point, but he may have forgotten to mention it. And I wanted all the subcontractor names in the initial quote so I can make sure we have all the lien releases before making the final payment, which shouldn't be a problem, right? Add that too. One last thing, it looks like there's a little double-charge for the dumpster rental, it's included in the job description of the base price but is also a line item price, do you see that? Let's just remove the line item, that will fix it. I'm making the decision tomorrow, so could you shoot the revised bid to me today? I know I said a bunch of things real fast, so I'm sending you an email with suggested text so it will take you no time at all. I wouldn't ask for this paperwork straightening up except you guys are the front runners. Thanks much."

Typo or deliberate, which one is true? Who cares! The end result is all the same: getting the best deal from the preferred contractor. Cleaning up vendor proposals is the most cost-effective time you'll ever spend. You'll never know what sour ending you averted. Misunderstandings eliminated. This doesn't mean there will be no callbacks, but you won't see flat-out refusals because it wasn't part of the deal.

Isn't an opportunist a swindler? Swindlers are unlikely to start off with a huge pricetag. They start the dishonesty with a low price to get your head in the noose. Most of the time they want cash up front, huge amounts of it. A super-low price in exchange for 50% in cash up front is not a good deal. Don't take that deal, ever.

A swindler may take a check. It's not the easy way to tell you're dealing with a swindler, that he won't take a check. Very seldom will they take a charge card, so if you have a personal policy of always paying the initial no-more-than-10% payment by charge card[7] you will have a no-deal situation with 99% of swindlers and possibly 20% of ethical businesses. That's a fine trade-off.

The Teardown Swindler has a plan, and that plan is to dive into teardown. Sometimes that's the only part they know how to do. While most swindlers insist or say 'it's the law' to get 30% or more up front, the teardown swindler will take a smaller down payment. When teardown has proceeded to a dangerous mess, they demand more money immediately. It

---

[7] Never use a debit card in contractor situations. If you're going to pay from a bank account, always write a paper check.

may be a perfectly normal-sounding reason. It may even seem to be entirely unrelated to the scary state of your house. Being over a barrel is only an impression. An exactly correct impression.

Victims often admit later they saw signs and hints that something was off about this deal, but they wanted the low price so much they ignored their inner barking dog.

Ghastly, awful stories abound about swindlers and home repair.

Case study: A homeowner hired a fellow to install a dormer. The contractor removed part of the roof, and then covered it with two huge tarps. He became Mr. Excuse on why he couldn't return: family funeral, car trouble and so on, for over a month. Materials arrived and were piled in the back yard. The homeowner called the contractor to report the quantity didn't seem sufficient to do the job, but he was assured all would be fine, more were coming.

Then, unbelievably, Mr. Excuse came to the house not to work but to 'borrow' the tarps for another job! As he did so he told the homeowner it wasn't a big deal because he was coming tomorrow to finish the dormer. He didn't, and a few days later a big storm water damaged almost every possession the poor guy owned. On top of that, the furnace conked out from overwork.

The homeowner called every half hour during the storm with no answer. The next day, the contractor's phone was disconnected. The homeowner finally called the police. When the swindler was caught, it turned out he wasn't a contractor at all, just an unemployed guy posing as one to get some cash.

Case study: the swindler was a guy who had once been a contractor but was on the verge of bankruptcy. He undercut the competition on a handful of major home additions, then wrote contracts requiring 50% due upon delivery of materials. In a few days he arranged to drop off scrap lumber and boxes containing rocks (empty except for rocks to prevent them from blowing away), got his payments and applied them to the mortgage of his million-dollar house where the money was safe from creditors. The homeowners were mystified by the deliveries but didn't grow alarmed for weeks, and by then it was too late to get their money back.

Swindlers wake up in the morning with the intention of taking your money and not doing the job. They leave you worse off than before. Your behavior, your gullibility, and your trusting nature do not make swindlers. You do not corrupt them or 'tempt' them. I don't make passersby into purse thieves by hanging my purse on the back of my chair, and you do not make swindlers by wanting home repairs for a low price.

With swindlers, you need to engage the authorities to stop them or else they will continue to perpetrate their evil-doings in your community. The authorities could be the State Attorney General's office, the District

Attorney, or the police. Murders and gang warfare are a tiny part of the police officer's job, even though TV makes it seem like that's all they do. Saving people in situations like this, anxiety-producing situations with swindlers and financial loss, give them something good to brag about to their spouse.

Every few weeks the TV news does a segment on a homeowner who lost a great deal of money to fellows like this, swindlers who find ways to tempt or con people with tricky or slick repair proposals or ways to make fast cash. By the time the police find out the fellows are long gone. It would be a happy day at the precinct if you phoned while the swindler was still on the premises. You might win an award. The police give those to people who help them catch swindlers.

Why do opportunists sometimes feel like swindlers?

Because overpaying results in a worse job. Not better. Not even equally as good. Worse.

There are several solid business reasons why the overpaying customer gets a raw deal. For one, the overpaying customer gets the B crew, not the A crew, since the job will make money even if there are screw-ups.

Two, when you overpay, the contractor will make the job take longer, sticking non-work days into the job to disguise the overcharge. He's trying to make it look like you're getting your money's worth but all you're getting is worse service.

Three, he may even outsource it entirely to some new contractor who is hungry for work but inexperienced, because he can do that and still make $500 or even $8,000 free and clear. This kind of outsourcing might happen to anybody, overpaying or not, but is far more likely to happen when you pay too much. When the truck that pulls up doesn't have the logo of the company you hired but that of some other company, it just happened to you.

Four, he is more likely to disrespect the overpaying customer, so will skip steps, use a staple gun instead of nails, install thinner sound-proofing insulation than promised, or whatever shortcut and cheap-out he can think of, since she's an easy mark.

Five, it's hard to justify giving the overpaying customer his best work because she can never become a reference. He can't risk her blurting out what she paid to a prospective customer and scaring them off, so her job is a backwater. Nobody can know about her. He's not going to put sufficient effort into making the edges align, removing blemishes, or whatever.

Lastly, with a large overpayment the holdback becomes immaterial. Costs are more than covered with the 85% to 90% already received. He may turn a deaf ear to requests for help with problems because he has a bubble of new work and spending more time on this job isn't a good use of his resources.

Overpaying an opportunistic contractor vs. a Swindler: Isn't that just splitting hairs? No, absolutely not! An opportunist is a normal businessman who pays Workman's Comp, donates to charities, has photos of his kids in his office and pays a designer for a nice company logo. You can brow-beat and complain and threaten to give bad reviews online to shape up an opportunist. You can even renegotiate the price after the job is done, as detailed in the chapter *Getting Your Money Back*. None of that will work with a swindler.

Opportunists are handled via escalating stages of repercussion; swindlers are handled by calling the Police and notifying the State Attorney General instantly, without giving the evil doer any warning.

It is vitally important to recognize which one you have. If you give a swindler some lip, it can go badly for you.

Highballing

There's a third type of overcharging that every customer needs to understand, only so they can play the game. It's called highballing. Highballing is providing a quote 30% to 500% larger than it should be because this business doesn't want the job. That's the key difference: he's not trying to make a killing, he's not trying to cheat, he's not rendering an opinion on the gullibility of the customer; he DOESN'T WANT THE JOB.

The timing could be bad or his schedule totally booked up. The job could be off the beaten path for his business or requires equipment he doesn't have or would have trouble getting. Perhaps there's poor access for his equipment or no parking for his truck. Maybe the job is too far away for a reasonable commute, or maybe he's had bad experiences with the supplier of the materials you want.

For whatever reason, he doesn't want it but he can't simply refuse to quote; his website might promise free quotes, or he told you on the phone that he would, or he's already arrived at the premises.

When your estimator asks if you're getting other quotes and doesn't seem happy if you say no, he's going to highball. When he appears to be encouraging you to find somebody else, he is. When you sense your estimator is reluctant, ask him for the names of some other shops that do this work. I've done this a half dozen times. Don't be surprised if his face lights up and he works a bit at coming up with some other names. If providing competitor's names and phone numbers doesn't let on he doesn't want the job, I don't know what would.

If you don't take the hint, he'll probably highball. Some guys can't admit being out of their comfort zone or feel disloyal recommending other shops, so you can get a highball quote without warning too.

Highballing is ten times more likely if you phone for a quote rather than use a written Request for Quote. Why? People find it harder to decline-to-quote to a person on the phone. A letter or email they'll simply ignore.

The intensity of a person asking for a quote is more overbearing, so there will be cases of the poor-fit vendor providing a quote in halfhearted fashion to a phone call request that he would have ignored in an email. Do you really want those quotes? No.

The problem with obtaining only three quotes is that unless you know a lot about this work, two of the three may not be good fits for the project. Three quotes are fine for professionals with their stable of steady subcontractors. Homeowners who don't do this every month need to send out 9 to 15 Requests for Quote to make sure they get at least five good ones. The initial selection of contractors based on tiny ads or skimpy websites may include a few who see this job as borderline to their scope even though it looks close.

When you're inexperienced at this, talking to only three contractors could mean that two of them highball. That makes the remaining quote look astoundingly low, when in fact it could be on the high side. It's unfortunately common to receive several highballs for inner city home repairs, so even after obtaining three quotes, a homeowner might pay four times the real cost!

Highballers who win the job usually subcontract it and still make big money. That wasn't their intention, but you backed them into a corner. You're better off soliciting ten bids to find the shop the highballers intend to hand it off to, and skip the middleman.

Contractor Heaven

If contractors wrote the rules, they would throw out a pie-in-the-sky price on every job they want to get and the homeowner would either pay it or balk. If they balk, he lowers it or considers their counteroffer. No customer would ever rule him out until they had a little back and forth about job expectations and price. Every bid he jogs up results in serious consideration, not silent tossing into the garbage can.

Contractor heaven is having the freedom to name his price but retaining the second option of a smaller profit margin and the third option of only breaking even on the job if he needs it to keep his guys working.

If it comes to it, he WANTS to match the price of his competitor, if only to keep the job away from his competitor.

Contractors don't write the rules. He can't get away with a pie in the sky price for everyone. Customers who know what it should cost will just discard that outrageous quote and not counteroffer. With every new customer, he looks for subtle signs that he's dealing with one of these. Depending upon his perceptiveness, 40% to 80% of customers get a fair price right off the bat.

Every now and then he perceives during the estimating process that he has a lock on the job so he can go pie-in-the-sky. Why not? It will counterbalance any break-even jobs he had to take.

Contractor heaven isn't about making a killing, it's about staying in business. The second and third options are part of contractor heaven too.

The more often we select reputable businesses and the less often we give money to businesses with deceptive practices or shoddy workmanship, the more the good guys thrive. I like the idea of the good guys thriving.

Every time a homeowner gives money to a swindler or incompetent guy who hung up his shingle in some trade, that's one less job the competent contractor gets. When he loses six jobs in a row by 5% to shops he knows leave unhappy customers in their wake, what can he do? If he tries to warn his potential customers, he's just bad-mouthing.

Good preparation before the estimator visit not only improves your outcome, it supports the right businesses and starves the baddie businesses.

Contractor heaven is about staying in business. The good, competent contractors can't survive if you don't hire them. I cringe when an acquaintance tells me they liked and trusted one contractor the most, but his price was slightly higher so they signed with another. Don't do that! Keep calling that best shop until the price works for you. In hindsight, explaining to friends that you picked this other contractor to save money is pretty ludicrous when your life is in a shambles like Mr. Hole-instead-of-dormer's situation.

Remember him? That poor fellow with the hole in his roof where the dormer was supposed to go? On his side of town I'll bet there are four firms that each built three dormers this year and would have loved to do a fourth. There's a worker in his state who has supervised the installation of twenty dormers. If he had found that guy, imagine how boring and uneventful his life would be with his new dormer. He'd gripe to friends "He was $1,000 more than the lowest-price guy."

If you pulled out your crystal ball and told him the fate that $1,000 saved him from, would he believe you? He might reply that's impossible, no one would do that to another human being.

Swindlers really don't care how much they ruin your life. Their view is if you are hoodwinkable, good. Protect yourself; don't save any Nigerian businessmen, don't send any money to stranded grandchildren inexplicably in Canada, don't let someone cut you a check for $1,000 more than the item you're selling and give them the difference, and don't hire some guy without a business street address or proper credentials who refuses to accept charge cards.[8]

The people who were ripped off by the verge-of-bankruptcy contractor because they paid their 50% upon delivery of 'materials' also did not do their setup work. A brief review would likely find poor BBB ratings, Angie's List complaints, and lawsuits on record.

In this case, they didn't even need to look anything up! Being required to make a 50% payment before any work was done should have been enough red flag for them to find a different contractor.

Contractor heaven is about rewarding decent contractor behaviors and punishing bad ones. Mr. Verge-of-bankruptcy was richly rewarded for being a corrupt, lying, stealing scumbag.

---

[8] Every contractor can accept charge cards. Paypal and Square One, among others, provide the card-running phone attachment for free plus allow simple type-in-the-card-number entry, have no monthly fees and deposit the money into his account. Some good contractors don't take cards only because they're not computer savvy and don't know how to set it up. This hurtle can be crossed in one day if he gets his accountant or a teenager to help him.

I don't care how big and wonderful your contractor used to be, when he asks for 30% or 50% before he has incurred 15% of the labor, just say no. Payment follows work. Materials? He has 30 to 45 days to pay without incurring interest, so payment for those can lag even farther behind than labor costs. All businesses should have a line of credit at a bank, so they absolutely can pay their suppliers in full prior to receiving your payment. They don't even have to tap the line of credit, they can do it with cash flow.

The less you know about the work you are hiring them to do, the more your deal must be nothing down, 80%-90% on completion with a 10% or $500 holdback, whichever is larger. Let him work like the dickens for every percent of concession from that stance. See the chapter *Payment Schedule* for the full details on this.

Whether contractors need to be licensed varies from state to state. Some states require a wide range of contractors to be licensed, some require only a few trades, and others have no licensing. In those states anyone can hang up a shingle as a contractor or GC. States that require licensing have developed statistics on the building trades, and those statistics are worrisome. We have no reason to believe the stats are better in non-licensing states.

Maryland is one state with a robust contractor testing and licensing program. Statistically, almost 50% of first-time contractors do not renew their licenses the next year. Only 10% remain in business more than five years.

The risk of your contractor going out of business before finishing the job is significant. He may not see it coming himself at the time you sign. If the worst happens, it will be a miserable experience. Don't make it worse by paying him in advance of work. Also, insist that a list of the major suppliers and all subcontractors is included in the contract before you sign, so you can call them to see if he's already deeply in arrears. Get that list!

Contractors abandon their own fledgling business for only one reason: debt. Nearly always, their last ditch efforts to keep the business afloat involve demanding large upfront payments, refusal to take credit cards, saying things to discourage the homeowner from obtaining other quotes, and accepting jobs that are on the fringes or entirely out of their skill set. You need to steer clear of any business showing one of those red flags.

Clean and Legal

I am a fan of above-board, clean and legal purchasing. When a buyer steps into the realm of under-the-table freebies, kickbacks, unlicensed, under-qualified workers and so on, all bets are off. Those deals work out well perhaps 20% of the time, maybe not even that much.

The other 80% of the time the buyer is too ashamed to tell anyone how sour it went towards the end. If he bragged about it in the beginning, he keeps up the farce. He is unable to enlist any legal or professional organization assistance due to his own culpability. So even when you think you know people who got a good deal by going outside of the law, you don't.

Offering to pay cash can provide a small discount but you lose the credit card protections that are described in the *Credit Card Court* chapter. Cash payments work only if nothing goes awry. Often what goes wrong is a turn of events you could never imagine. There is less financial risk if you keep things in writing with hard-copy receipts when you are in unfamiliar territory. This is the bare minimum necessary to enlist the law or the professional review process to intercede on your behalf.

The big plus to keeping the agreement in writing is that it prevents problems. The recourse part is a pain in the neck. When the other guy sees you are a get-it-in-writing kind of person, your project goes far more smoothly than if he felt for a moment he could get away with pretending it wasn't part of the agreement.

The perfect contract keeps you out of court and ends with an outcome that meets expectations for the agreed-upon price.

Any layperson can write a perfect contract by using the advice in the *Contracts* section.

Second Quote Phenomenon

After obtaining hundreds of competitive quotes in at least forty industries in several states, I have discovered the second quote phenomenon is a universal truth. It applies when buying services, repairs or custom-made items regardless of the size of the purchase. It applies whether the sellers solicited are Fortune 500 companies or one-man shops.

The second quote phenomenon is that the second quote, and all ensuing quotes, are overwhelmingly likely to be lower than the first quote. The first quote is an animal all by itself. The first quote is so frequently opportunist, or the wrong vendor, or misses an important value-adding aspect which turns out to be critical that I affectionately call it 'the discard.' The exception to the first quote never being the lowest is when by some accident the first quote is from a swindler. Then the price is a pretend-low price because they are out to rip you off.

The first quote is not slightly more likely to be the highest, but is the highest bid more than 50% of the time. If you get five quotes, the first will be highest 50% of the time. If you get fifteen quotes, the first will be highest 50% of the time. As you greet successive estimators knowing more and more, the price of the legitimate businesses will descend until it meets time and materials. Not in a perfect line, but in a trend.

The likelihood of the first quote to be the lowest of the good quotes when obtaining is something I can't pin a statistic on since I don't recall it happening. That isn't to say it never happens, it just has not happened to me in the past ten years.

People say to me "Why should I get multiple quotes? That's too much work." It is such a slam-dunk price reduction to get a second quote that it's incomprehensible any sensible person would fight it. Most of the second through eighth quotes are so overwhelmingly likely to be lower than the first that the payback on time spent is something between $100 to $2,000 per hour—it depends upon the price of the job. Each one takes far less time than the first did.

You are going to learn a lot from the first guy. This could be why he's highest, out of a feeling that he should be compensated for spending so much time teaching you about his field.

Use this learn-most/highest-quote tendency to your advantage. Select a farther-away shop or one you are only lukewarm about for the first site visit or discussion phone call. This way the learning curve is climbed before you speak to your preferred shop.

Speak to your preferred shop second, third, or better yet, fourth. Having even one other estimator in before contacting your favorite contractor does wonders for getting the best price from him.

There is a tendency to make the first estimator you speak to become the yardstick and measure all the others against him. Resist this urge.

You'll learn more from each estimator you speak to. But be prepared; after speaking with three or four candidates, no matter what type of project, you'll find they disagree on something fundamental. One recommends something, and another says that should never be done. One says C is best, another says T is best. Each one will sound absolutely certain. Use the internet to figure out the truth. Never assume the first guy is right and the later one is wrong. Keep asking questions, because beyond a doubt one guy is more right.

Because opportunism is strongest in the first estimator, he is statistically more likely to pad the quote with unnecessary extras, exaggerate the amount of work involved, tell you a middle-quality product is top of the line, or misrepresent something. Be extra skeptical of the first guy if his 'must do' is also significantly more expensive. Had the first guy been the third or fourth estimator, it would be obvious he was being opportunistic.

The first contractor, while you learn the most from him and get your first feel for their business from him, may or may not be the most honest, highest quality contractor in the bunch. When later estimators indicate they can do the job in half the time, don't assume it's due to cutting corners or rushing things. It might be, but it's statistically more likely that the first guy padded the quote.

As for that overly-high first quote, the guy is a not lost cause. If you favor a bidder but his price is too high, don't just rule him out. Give him some Contractor Heaven. Call him back and let him know the price to match or beat, after confirming the specifics are identical. Make sure he isn't higher because he's using higher quality components or is including extra work. Ask about areas where his quote may differ from the other ones. Will you permit him to use the lower cost materials to match the lower quote?

Lastly, you might be tempted to ask him to match the lowest quote which came from a guy who gave you the creeps and you'd never hire. Don't go opportunist on your favorite guy. See what a slippery slope that is? If you're asking him to match a price, be sure it is really the firm that will get the job if your favorite guy says sorry, he can't match it.

Finding the Contractor

The less you know about it, the more bids you need to solicit.

Getting three quotes is for General Contractors who have their list of subcontractors they know and have worked with before. Getting three quotes is only for people in the business of hiring contractors.

You don't have a stable of regulars. You may not be certain every one of the companies you solicit has done a project of your size with the materials you want. Because of that, cast a wide net; their proposals and their prices will form a pattern, and after they answer a few questions that will arise in your mind, you will be able to pick the one with experience in your kind of work.

Look around your area in a forty mile radius: there are hundreds if not thousands of houses with the thing you want. People installed every single one. Experienced workmen exist, dozens of them. There is absolutely no reason to hire a first timer, or someone whose skills are close but no cigar. If you want a decorative patio, why hire a totally uninspired guy who does sidewalks for the city? He's going to give you a slab. There must be ten guys who install lovely patios in your area code. Hire one of those.

Pay attention to job size and age of your house. It is bad to hire a guy to build a sunroom who, up until now, has only installed windows, but it is also just as bad to hire a guy to replace a window in a 1920s house who, up until now, has only done brand new construction.

A guy with 'kinda-close' skills in bigger, smaller, newer, older, harder, softer, and so on is not your guy. This isn't a DIY book, so I'm not going to deep dive into all the variables between new construction and retrofitting into an older home. The differences between working on a home built in 1992 and one built in 1896 or even 1947 are profound. Even if you can't imagine what they are, respect the workers who specialize in your project in houses of your age and construction. Experience in the age of this house means the guy can see through walls. Figuratively.

Construction workers do not learn their trade like you learned math. When you learned math, each chapter in the math book covered a concept and then you were asked to solve eight word problems using variations of the formula. Using the same analogy, construction workers learn by simply plunging into one of the word problems under the eye of an experienced worker, with no variations on a theme, and often no explanation regarding the steps performed. Concept is not mentioned.

When done, the worker knows just that one way, with those variables, and not a drop more. He or she will learn more conditions the second time, and more the third. If supervised by an experienced worker through several problems, great! They will be guided and get a good outcome. In time, after solving perhaps five variations successfully, the concept may form in their own head. If that happens, handling problems beyond the examples seen thus far could happen. It may take twenty jobs to encounter some of the less-frequent variations. But often, new workers are shown once, and the second time they are on their own.

Having done it once before is insufficient experience.

If he's done it only once before, there will be variables on your job that he hasn't encountered before. Only experience can get him past those. Not book learning. Not native intelligence. Only experience.

Here is the politically-incorrect truth: the boots-on-ground workers in construction are not the brightest bulbs. Yes, there are exceptions—college kids with summer jobs, computer programmers who couldn't bear to sit and push keys for one more day, immigrant pharmacists who need to make money until they get licensed . . . but for the most part their mommas didn't consider them to be college material.

Most of these not-bright bulbs, however, can do amazing work after they've done it twenty times. They can run circles around a first-timer. If I have a choice between a 21-year old who has been working as an assistant to a 56-year-old tile layer for two years and a college kid who laid his first tile three weeks ago, I don't care if the college kid is from Harvard, give me the experienced assistant.

That is why you should not be overly impressed by a smart as a whip estimator, unless he will be on your site supervising your crew or even doing all the work himself. If he is, then by all means be impressed. If you like his ideas, his quick-on-the-draw mind, make it a contractual requirement he be on your site during the work.

Because going out of business during your project is a risk, finding someone who has been in business for several years shows they are doing their paperwork and managing cash flow far better than average. An easy way to find established contractors is to find an old phone book, five years old or better, and start calling. If someone answers the phone under the same company name, that's half the battle. The other half is checking their license with the state, checking their listing with the BBB, with Angie's List, and online reviews.

Cluster of Bids

You may have heard the often-repeated advice of tossing out the lowest bid. If you only obtain three quotes, you are discarding 33% of your quotes for no good reason. That doesn't make sense. It makes sense to toss the low bid out when you have more than six quotes and one is all by his lonesome 8% or more lower than the next lowest bid. Toss out two, even, if those two bids are 10% lower than the main cluster, especially if the main cluster is close.

After the bids come in, five to twelve of them, if they are all over the map, say 50% variation from high to low, something is wrong. Either different information was given to each bidder, or too much was left open to interpretation, or you have a smattering of gross opportunists and heartless swindlers. You will have to talk to estimators on the phone, send photos, or do something to reduce the apples-to-oranges state that your bids are in. DO NOT just pick one. When you have eight bids and cannot see a clear, fairly tight cluster of three to six of them, what you have is apples-to-oranges and you must ask them for details on what's included, or pick the materials, processes, size, color and so on that you like best and have them all quote those specifications, so a cluster appears.

If the lowest bidder on a $15,000 project is $500 less than three of the others, that is not suspicious, that's well within normal estimating variation, only 3% lower. Keep him in the pool.

Using a Handyman

In every area of the country, there's a lay of the land in the trades. When you don't know it, either because you're new to the area or don't know any construction workers, one option is to hire a good local handyman to help sort through the quotes. Selecting a handyman you like and trust is not stressful; you just phone him. They nearly always charge by the hour, are good at estimating their time, and if you don't like him, just hire a different one the next time. Using an older phone book is the easy way to get one who has been doing this for a few years.

The advantage of reviewing your pool of potential bidders for a big project with a handyman is that he knows the lay of the land—and sometimes literally, as in knowing what kind of ground prep is needed for home additions or garages based on the soil composition in your neighborhood. He's the guy who is hired when contractors mess up so he's likely to know all the bad ones.

On page 48 is a list of tasks and jobs that one might call a handyman to perform. If the job you're considering is on the list, try a handyman first, rather than a new-construction or large project firm. If you simply don't know, it can't hurt to send a few local handymen one of the RFQ or simply phone to set up a site visit.

What Multiple Bids Will Do For You

Case Study: If this is your twelfth deck in three years, then yes, you know a lot about the strengths and weaknesses of deck builders in your area. If this is your first deck installation ever, how can you know that Bates has a low price but uses cheap materials that degrade after three years, plus always installs the ankle-twister 11" tread stairs so should only be considered for decks with no stairs? You won't know that Grady talks everyone into one of those octagonal areas in the corner, and Martin's website says they do decks but they really only do screened-in porches these days. Then there's Bean, who prefers new construction and hates tearing out the old deck or patio so charges an arm and a leg to do it. Well, you can't know any of this as a first timer.

Even in something as simple as getting a deck built, to be happy with the result you have to cast a wide net to scoop in the vendors whose preferred deck style matches your desires. Say I get quotes from the four above plus two more. Bean brags about the 6" rise, 15" tread on his stairs and how it reduces fall injuries and twisted ankles by 90%. Grady presses you to add one of those octagonal things with built-in planters and comes in with the highest price. Martin tries to talk you into making it a screened porch instead, pointing out how the midday sun will reduce the usability of a plain deck, and when you decline, their quote never arrives. Bates has the lowest price but the stairs look really cheap.

You go back to Bean and say, "Can you edit the quote to include some built-in planters? Can you install one of those retractable sun awnings on the side of the house? I'm considering Bates for this job too, so could you sharpen your pencil on the final number?" Each bidder gave you something to think about and what you end up buying is a combination of good ideas.

If you call just one and hire them to build whatever they propose, what you end up with is not as useful and pleasing. Due to opportunism, the price tag will be higher. Try contractor heaven instead.

To get to Contractor Heaven:

1) hunt for several contractors who specialize in the work

2) develop a Request for Quote, RFQ, or just jot notes on all your requirements & dealbreakers so you don't forget to mention each one to all estimators.

3) Decide who your favorite is; talk to him fourth, third or second.

3) Mail, email, fax or just call with your RFQ details.

4) Review bids, weed some out.

5) Cherry-pick the best features, methods, and materials. Call the finalists back with the changes.

6) Tweak until quotes are apples-to-apples, not apples-to-oranges.

7) Pick the one you like best—NOT necessarily the lowest price. Then make his price good too, using contractor heaven.

This really isn't a lot of time; some of these steps take just a few minutes. Step one to three might be popping to the professional organization website, looking at the state's list of licensed contractors or checking the phone book, then jotting some notes and making a few calls. Provide your email for the completed quotes, look them over, call to change one or two things, then pick one. It flows naturally.

Online, search for their professional organizations or trade associations. Angie's List, Yelp.com and other customer-review sites are also good. At the end of this chapter are some, not all, of contractor's professional organizations.

Ask a Building Inspector which contractors routinely get their work approved on the first visit. If there is a lumberyard nearby, chat up the manager to see if he can provide the names of firms who pay their bills on time and buy the better quality materials.

Beware! A beautiful website doesn't mean they're legit. Anyone with $500 and a phone can get a good website. Send an RFQ based on looking over their website, but if they make the finalists, drive to the office during business hours. That is the minimum due diligence.

Lone tradesmen like an electrician may work out of a home office in the basement, but then you want to see scratched file cabinets, a filthy carpet and stacks of spare materials, things that indicate he does real work for a living. Something that looks unused and without parts and boxes is suspicious.

Read reviews carefully. Most of us can pick out when they were written by the same person using different names.

I looked at one vendor with over 50 good reviews. Every review talked about how nice the estimator and workers were, by name. Not a peep about the work itself. Fakey! Another had six reviews, and three of them contained the same word that was misspelled the same way. Fakey! Legit reviews often do have misspelled words, bad grammar, and call things by the wrong name, but each in their own unique way. A third vendor had seven nice reviews. Each review was three sentences long. What are the odds of that?

Professional organizations

Competent contractors often belong to professional organizations. Contractors choose to pay dues to join because of group insurance offered by that organization, newsletters and notices regarding legislation which affects their trade, authorized use of the logo on stationery and cards, and because people like you use it to find good contractors.

Contractors who join a professional organization care about their name, are more likely to pull permits, more likely to be conscientious about following building code, and more likely to see themselves as part of a trade and tradition.

Paying the dues and using the logo doesn't mean 100% of them are good, nor does it mean that a shop that doesn't join is bad. Some guys think of themselves as 'not joiners' and can't seem to get past that, even for business reasons.

URL	Description
http://www.handymanassociation.org/	ACHP Handyman
http://theuha.net/	United Handyman Association
http://www.phccweb.org/	Plumbers and HVAC
http://www.caphcc.org/	Plumbers, California
http://www.necanet.org/	Electricians
http://www.ieci.org/	Electricians
http://www.findapainter.com/	Painters
http://www.concrete.org/	Concrete
http://www.nrmca.org/	Concrete
http://www.ascconline.org/	Concrete
http://www.csia.org/	Chimney Sweeps
http://www.masonryinstitute.org/	Masonry work
http://www.finishingcontractors.org	Drywall and adding basement or attic rooms
https://www.nahb.org/	Home Builders and additions
https://apsp.org/	Pools and Spas

Landscapers have state organizations; search "[state] landscaping professional organizations"

URL	Description
http://www.nrca.net/	Roofing professionals
http://www.aloa.org/	Locksmiths, security systems
http://www.sopl.us/	Locksmiths, security systems

Handyman or Contractor?

When something goes amiss on your property, if you call a contractor specializing in installing new, his answer will always be to tear it all out and put a new one in.

I want to say this twenty times real loud: guys who sell new ones will tell you to replace it. That doesn't make it the right answer; it's simply the answer that puts money in his pocket. He isn't a bad guy for doing this. What he installs may be robust and have a ten year warranty. But it diminishes your savings and possibly throws you into debt for a year, robbing you of pleasurable vacations, dining out, and other luxuries. It doesn't improve your lifestyle. Do this six times in nine years and you'll feel like the only reason you live is to feed the house monster.

When something could be repaired and you ask a sell-new guy for a professional opinion, this is what you are saying: "Would you like a 3-hour job at $60 an hour, or would you like a 16-hour job with a bonus $800 markup on materials?" He gets to pick. Hmmm, toughie . . .

For many issues that arise in a home or yard, first phone someone who does repairs. A reputable guy guarantees his work for a year, so if it truly can't be saved it's likely he will tell you. I'm not saying hand out trust like candy at a parade; no one in construction is truthful 100% of the time, I'm saying your odds are much, much better for an honest appraisal of the situation if you get a trustworthy repair guy in first and then follow his recommendation. If he's honest enough to say any repair he did wouldn't last nine months so get it replaced, pay him for the consult; this guy is a keeper.

Calling a roofer for a leak, a window salesman when the rotating knob breaks or a house painter when the corner of the house gets scraped is how people diminish their whole lives and never get to take the kids to DisneyWorld. Constantly replacing expensive things that could be repaired for $1/30^{th}$ of the cost is the way to be in debt for no good reason.

Repair is not second best. Repairing a rotting windowsill with mold-killer and the wood sealant made for that purpose, for instance, makes it far more resistant to future rot than a new window. More to the point, putting off replacement for six years means less lifetime cost if you live here for fifty years, and even less if you move away in four. Besides, in a few years you might remodel or simply want a different color or look.

One common sales pitch you'll hear is, "it will only last three more years, so might as well replace it now." That is jaw-dropping stupid. What if I said about your shoes, "they will only last more three years so throw them away today." What if I said about your ink pen, "it's 80% empty so throw it away today." Your reply would be, 'well, I'll use it until that day comes, I'm not made of money, I'm not about to throw away things that are still doing the job." That's the proper attitude!

When you have a leak in a 15-year-old roof, when the caulking has dried out and cracked on a 20-year old skylight so it drips when it rains, when the wood floor has several ugly scratches, when a couple of pieces of siding are damaged but you have a few spares in the basement, when several cracks in the driveway are opening up, call a handyman.

Picking a handyman should be done the same way as any other contractor, by talking to several and finding the most experienced one at this particular thing. No matter what their ad or website says, no handyman does everything; every one of them has tasks they do great, and others not so much. One may be great at fences, installing shelves, hanging curtain rods and patching asphalt, but be afraid of heights so won't do roof work. Some are better with wood, and some with motors and things like installing a replacement dishwasher or a garage door opener.

For that reason, it's worthwhile to talk to several handymen in your area about recent projects and their personal strengths. Take notes on their skills, so later on you can call the appropriate one right off the bat.

Handymen have usually lived in the area for many years, so are a fount of information on local contractors. One good use of a handyman is to ask him who is good and who to avoid when starting a major project. Pay him for his time, just like a consultant. His advice will save you many times more than his fee.

On the next page is a brief and incomplete list of tasks that handyman can tackle; when you need any of these, seek out local handymen first, not contractors specializing in new. As always, you never want a first-timer to work on your house, ever. Make sure he has done this before.

List of handyman jobs:

AC, Put in/remove window and wall AC units

Doors, fix or replace

Doorknobs and handles, install or repair

Dead bolts/security locks on doors & windows

Install pet doors

Photography, monitoring, insurance documenting

Closet organizers, move/install new poles, components

Windows, repair /install screens, storms, clean outside

Window repair- rot, loose, stuck

Cabinet repair, hinges, mounting

Patch holes in wall

Gutters, clean, repair, install

Downspouts, clean, secure, fix

Drywall and plaster wall repairs

Painting, interior

Toilets, fix leaky/noisy toilets

Faucets, fix leaks

Faucets, shower heads, install new

Unclog drains

Kitchen, countertop repairs

Install new appliances, remove old ones

Recaulk or regrout bathrooms and backsplashes

Ceramic tile repair

Wall mirrors, install or remove

Hanging curtains or blinds

Fireplace cleaning, brick cleaning

Fence repair, installation

Deck repair, staining, installing

Asphalt repair / concrete patching

Driveways, sealing, patching

Porch repair: steps, railing

Senior living modifications

Organize basements and garages

Pegboard, shelving and cabinet installing

Roof, caulk leaks around skylights and stack pipes

Roof shingles, fasten down or replace

Skylights, install or repair

Solar panel mounting
Satellite dish mounting
Flower boxes, mount or remove
Flowerbed trim, edging, shovel work
Landscaping, tree trimming, planting
Sprinkler system installation, repair
Swimming pool maintenance, pump replacement
Shelves, install
Wall molding, chair rails, wainscoting, install, fix
Run phone, cable or ethernet lines through the walls
Plug outlets and light switches, replace
Dimmer switches or GFI outlets, install
Garage door openers, install & replace
Dryer vent cleaning / installation
Pool maintenance
Sump pump, replace, repair
Repair water damage/ stains
Wall mount TVs, install or remove
Light fixtures, replace
Light bulbs, replace hard-to-reach ones
Bathroom exhaust fans, repair/ replace
Ceiling fans, repair / replace
Ceiling leak/damage repair
Wood trim, furniture: sand and stain
Wood floor sanding, polishing, repair
Wood furniture, repair, staining
Attic insulation, lay in place
Chain saw, tasks involving one
Shrubs: remove dead, plant new
Set up or remove Christmas decorations
Help with moving furniture or boxes; junk removal
Assemble sheds and furniture
Install or remove things requiring measuring, sawing, drilling, screwdrivers, pliers and arm strength.

Many will do things that simply require arm strength and time, such as moving furniture, cleaning windows or digging holes.

When you have a handyman you trust, it's not a waste of money to have him supervise other trades doing work on your property. Having a knowledgeable person looking in and dealing with issues in-process can keep the project on track and timely.

In addition to projects, some handymen might agree to keep an eye on a house while you are gone, including watering plants, mowing, feeding caged pets, checking for weather-related damage, doing pool maintenance and other tasks a homeowner usually does themselves when they are home. It's a safe assumption a handyman will do any task within his abilities that you're willing to pay him to do.

The Estimator Visit

Very few home improvement or home repair and replacement situations can be quoted sight unseen. This means someone from the company will need to take a look. This person is the estimator.

That may not be an actual title. The estimator is someone with experience and some ability to calculate time estimates—not always, but usually. They take measurements with a tape measure and check for problem areas or things that would increase the cost like poor access, damage, any one of dozens of things that detract from perfect simplicity. No job has perfect simplicity.

The estimator may be the President of the company, or a Supervisor, crew lead, worker, or someone hired just to estimate. Estimators aren't salesmen, although they read the same books and might adopt some of the same annoying and counterproductive mannerisms of salesmen.

Do not rule out a contractor company because the person who shows up at your door has dirty shoes, unkempt hair, or looks and smells like he's been working all day. Neat and clean is somewhat suspicious with these folks.

I'm not saying a grubby appearance means he's good at his job, or bad at it, I'm just saying it should not be a factor in whether you rule him out. Make the decision based on his apparent knowledge of his craft, if he has a plan and shares it with you, if he seems smart, conscientious and pays attention to what you say and responds to it.

Few people are worse at reading people than I am. Because of this, my decisions are based on the content of what he says, and not any airs or sales patter or technique he may be using.

Having a realistic evaluation of my people-reading skills means I never hire someone on the spot. I continue with the plan to talk to several contractors. Later, in hindsight, it can become apparent that the phrase or quip which prompted the feeling of trust was just some rehearsed gambit.

I believe there is such a thing as being a good judge of character, but I was born without it. My experience is that 90% of people are born without it, but 50% of those behave as if they had it. Things go well or go bad randomly for them, but when they go bad they consider it an anomaly.

I cannot teach you how to be a good judge of character when hiring contractors because I can't do it myself. I can, however, teach you how to hire good contractors when you possess no ability at all to discern their inner soul.

There are three types of bad contractors that are easy to spot without sophisticated interpersonal skills: one, the guy who is just too sloppy and lazy in everything. His car is a mess, his clothes don't fit, his paperwork is awful, and when chatting with him he tells stories where his lines are 'Whatever, dude' and 'so I just left.' His answer to everything is slow down and do less. If tangible evidence of conscientiousness in *something* isn't there, I don't hire. You might think, well he's not doing the work, he's just doing the estimating. Well, the company made him their front man, so that says something about their work ethic.

Two, Mr. Slick. Often a fast talker, but could also be just a smiley slow-speaking guy. He wants to give me price breaks or throw in upgrades because he likes me fifteen minutes after meeting me. If people tend to hand you money because they like you, then it could be real. Otherwise it's a ploy. Think hard about hiring a guy who sells using ploys. It's easy to be impressed at the time, but three hours later realize he adroitly side-stepped a major question. My urge is to phone him to ask the question again; perhaps he just got carried away with a thought and went off on a tangent. If the question was a very basic one, and he pretended I didn't ask it and talked about something else, then it can't turn out well for me. A contractor may not be able to answer some questions off the cuff, but he should at least promise to call me tomorrow with the answer. With this guy, the conversation always goes in the direction he wants it to go. I talk, but somehow have no effect. He's too slick, too tricky, too well-rehearsed. He thinks of me as a mark. He's off the list.

Three, Mr. No-Plan. When asked how he will tackle it, how many people will be working or what equipment they will use, he won't say. He will say something like "We'll get it done" or "it's not a problem" in lieu of talking about the process, which should actually be running through his head at this very moment. I can't tell if he has no plan but hopes one materializes later after he gets the job, or if he's so cagey and secretive about his work habits that it borders on an attitude or emotional problem. It doesn't matter which it is, I don't want to deal with it. I never hire Mr. No-Plan. Is that a good decision? I don't know, simply because I never take the risk of hiring him.

One thing I try to overlook is the guy Who Doesn't Return My Calls. Not after the contract is signed, which is bad, but when trying to get a quote. This means the guy is busy, which can indicate he is good. I keep phoning and trying to reach him through my site visit process. Unlike other businesses where people at work are in their offices, by definition a contractor at work is off somewhere working with his hands with his phone turned off. No one likes their contractor stopping work to talk on the phone every fifteen minutes. If he isn't taking calls for three or four hours in a row, he's doing it right. I'm not so hot on Mr. Always-Picks-Up-the-Phone. Now maybe he doesn't return them because he doesn't want my job. That also doesn't mean he's bad, it means he's picky. He'll only do things he does well. That makes him an excellent choice if my job is his kind of work.

Another thing that doesn't bother me is Mr. Swears-A-Lot. It's just a bad habit with them, and has often even ceased to indicate anger at something. Blue collar people just pepper them in like meaningless extra syllables between words. They enjoy becoming wildly shocked and apologetic if they let fly when a woman is nearby. They like to think they have a 'just us guys' language and another language for when women are around. I accept his apology as often as he gives it, which is often.

This chapter is placed before the chapter on writing a Request for Quote, or RFQ, because household repairs are often initiated by phoning some likely candidates and setting up a time for the site visit.

The handling of the estimator visit follows the same guidelines whether you have phoned or sent them an RFQ. A homeowner may find themselves calling repairman, handymen, or installers twice a year, while tackling a project better served by an RFQ once every three years. On average, that is.

To put it another way, the handling of the estimator visit is vital and critical to getting the best price and good work regardless of the way the business was contacted. The RFQ is an efficient and practical means to handle bigger projects, but isn't as critical as the estimator visit. As you'll see, if you do the RFQ process right, you'll be amending it as you go along. Mistakes in the RFQ aren't a big deal because you'll fix them before awarding the contract.

Putting aside swindlers and unethical businesses for a moment, estimators are there to determine four things.

One, can you afford it? Being poor is a risk factor.[9] Poor people do not get mercy price breaks, they get highballed because the estimator decides it's too risky. He makes this decision based on what you say and do, plus how well you take care of your stuff. Begging for a low price because you don't have much money is likely to backfire.

He may ask you if you have a budget; with very few exceptions, the point beyond which you would be financially ruined is none of his beeswax. If you slip up and tell him, he's going to make an effort to collect every dime of that budget no matter how much you downsize your project. Step downs

---

[9] For details, see *Addendum B, Ten Risk Factors* that increase the price.

in material quality will have almost no effect on the total price. Your budget is the amount you just said he could have. If in your mind that figure included appliances and curtains with the kitchen remodel, or included shrubbery as well as the deck, well now it no longer does.

Two, is there any other reason he doesn't want this job? If the fellow decides this, the visit is going to get odd and uncomfortable. He may highball or tell you fantastical stories. He's doing this to become unappealing to you. You should hope and wish he just highballs. What might happen instead is the fantastical story. The hapless homeowner tries to integrate this baloney into the job with the other bidders and things get ridiculous. I'd like to describe this better, but all baloney is different.

At the risk of doing you a disservice, because this is not one tenth of one percent of the nonsense . . .a roofer might say the roof angle is illegal and the whole roof must be removed. A drywaller might tell you the house will fall down if that wall is removed ('support walls' merely mean a beam or two must remain), or to fix the gurgling noise all the plumbing in the house must be replaced. When you speak to six estimators you can clearly see the odd man out, but when you have only one or two you won't know what to believe.

One of the ways to discern the above is happening is if he launches into the major-huge-work description before pulling out his tape measure. The usual way, when there is some huge bad thing the homeowner isn't going to like involved, is to take measurements first, ask the homeowner questions about the job scope, then break the news slowly.

Three, he's there to take measurements, determine access, assess the condition of the building, figure out where to store things, hear your vision of the outcome, and assign times to labor elements.

Four, he's there to convince the homeowner his shop is the best one for the job. This will be the longest part of the visit. The more questions you ask, the better feel you get for his shop's usual way of handling things.

Available Time for the Site Visit

What if you send out eight RFQ and all eight want a site visit but you only have time to do four? Do some triage at this point. Or nest them closely – "Thursday at 4 is great, but we must be done by 4:45. If you arrive late I'm afraid the end time is still 4:45. Please call to reschedule if you are running late." He will know that the next estimator is coming at 5 PM. Doing this will have no negative impact on the caliber of the quote and may even improve it. Often, the estimator is mulling in his head how long the visit should last: too short and the customer might feel breezed by, too long and the customer might get uncomfortable. Now he knows and can meter his activity, certain he is meeting the customer's attention needs. For the majority of home projects, an experienced guy can see all he needs to see in ten minutes to generate a quote; the rest is the pitch.

Often people schedule estimators back to back like this and then act odd for half the visit, trying to rush him out the door. Please don't do that. When he arrives, shake his hand and tell him "Thanks for coming, I hope we can cover everything you need before 4:45, We'll have to cut it short at that time." If he starts getting long-winded around 4:35, interject "Do you think we're on track for finishing in ten minutes?"

Jargon

When speaking with you, contractors and skilled tradesmen will call it what they always call it, even if you have no idea what that word entails or signifies. They'll use jargon. Much of industry jargon consists of common everyday words with a different meaning in that trade. In the skilled trades the way the words are strung together can sound grammatically incorrect but are actually the correct terminology.

When speaking to a tradesman, there's a risk you'll talk in plain English and the tradesman will hear jargon, or vice versa. I can laugh now but it wasn't funny at the time, the day I wanted a new haircut and tried describing it to the hairdresser. Kind of stepped in the back, I might have said. "You mean tapered?" she asked. Sure, I replied. She cut the all the hair on the back of my head less than 1" long. My scalp was showing. Aghast, I pointed to another lady saying "Like that in the back! That's what I wanted!" The hairdresser sniffed, "That's layered. You said tapered."

To the ordinary ear there is little difference, but in the jargon world there can be miles of difference – or in the case of my haircut, at least two inches.

Tradesmen know what most people want, so they are used to filling in the blanks when a customer doesn't quite know what he wants or can't describe it in technical terms. But that may not be what you want this time, so having pictures or samples helps.

I've worked with tradesmen for over thirty years, and guarantee that using their jargon, meaning the correct terms for key parts, is the most profound thing you can do to show respect for their field. Calling everything a 'thingamajig' might seem cute to you, but to them sounds like an aversion to learning the right words for their craft. It is not presumptuous to attempt to use the trade term for the feature you are discussing.

While preparing to travel to Paris, a friend gave me a bit of advice for getting around: if you speak in English, they will often pretend to not understand; however, if you make a stab at speaking French, no matter how inept, Parisians will draw up whatever English they have in reciprocation. I found this to be exactly true in Paris, and it's true of tradesmen and their jargon too.

When my heat pump HVAC unit needed to be replaced, I doubt if anyone could be dumber about furnaces. Initially I called the two parts 'furnace' and 'outside part.' In the trade they are called 'heat pump' and 'fan coil' respectively. Before looking into it I guessed the outside part is the heat pump, because it has a compressor, or condenser or something pumpy in it, while the inside part with the air returns and blower probably has a fan in there. There are also two kinds of refrigerant, which I started out calling 'coolant', R22 and the new one R410a.

After the first contractor came, I read up online, becoming familiar with R410a, refrigerant, heat pump and fan coil and used the words properly in a sentence with my estimators. The only high quote received was from the first guy who took a look before I learned my new words. He was $1,200 more than the rest.

Getting an Outstanding Deal

There are three things business value more than money—more than money on a single job, that is.

The three things are: Level scheduling, Easy work, and Future Business.

The customer who provides one of these things can expect the company to do the work for a smaller profit margin. If the customer offers two of them, this one job can be done at break even. If you can touch all three bases, it wouldn't be unreasonable to have this first job done for cost of materials.

A company will give you standard pricing, however, unless you clearly offer these and ask for your discount.

Level scheduling

The smaller the business or the shorter the contract length, the harder it is to keep all their full-time employees gainfully occupied. Work tends to come in surges, the proverbial feast and famine. Uneven work schedules means paying employees for parts or whole days when there's no productive work, and when there's a surge in work, paying them overtime. Mandatory overtime to cover peak times causes burnout, increases injury levels and makes people grumpy.

Companies love moving peak work that requires overtime and moving it to slow times when workers don't have enough to do. Shops often devise off-season sales and specials, doing the work for 20% less. Even if a company doesn't have an official off-season discount, finding out their slow time and offering to plop your job right there if it might mean a price cut for you could make them glad to dream one up.

Another way to help companies level schedule is being more flexible. A local tree trimming service offers a 10% discount if the customer agrees to an 'anytime in the next three months' contract. This way he can fit the job in when his guys are nearby or on a slow day. There might be other businesses that are willing to give a discount for a flexible start date; it's worth asking about.

If there is flexibility, very early in the conversation I'll say, 'you know, I don't want you turning away work for this job, I'm in no rush.' Or I let them know they can break it up or stretch it out, using my job as a schedule filler. I'll ask if they have a slow period.

Another way to find out if you can gain a level scheduling discount is simply shooting the breeze and asking about busy and slow times. Estimators will usually consider it easy small talk. Then you can evaluate if holding off until the slow time is a possibility, or if you'd rather not. If yes, then open with "Does your company have a coupon/deal/discount for moving something from peak to off-peak?" Make it clear this isn't just you rescheduling for your reasons, but that you are trying to help them.

Easy work

Easy work doesn't mean physically easy work or simple work. It means work this shop frequently does, something well-aligned with their strengths. With their bread-and-butter work, estimates are accurate and require no just-in-case money. If yours is an easy job, when the estimator arrives tell him yours is an easy job and point out why – good access, mechanical assistance available, near to the office, low to no removal costs, real similar to six other things you know they've done recently, whatever applies.

By framing your job as an easy job, of course only if it's true, your estimator will mentally fill in the blanks with lower cost labor and smaller transportation costs and so on.

Naturally, the estimator doesn't need your help to see that it's an easy job. Showing that YOU know it's an easy job can double or quadruple your chances of seeing a lower price quote. Do this as early as possible; consider opening the door, greeting him, they saying, "Hi, I'm Lynn, this is an easy job for you guys, let's go over here to take a look." Reasons it might be an easy job for them include: ads mentioning they own special equipment for this, your house is very near their office, or they did the same work on several houses in your neighborhood.

Most estimators work from a list of standard elements for a job. Their standard work-up may allow for a half hour of travel time but you're three minutes away; asking for a break because of this could lower a service call by $60. This works best if you also ask to be either the first or the last visit of the day. Say this job doesn't require framing, some customary preliminary work, or no disposal costs. Be sure to ask for these customary elements to be removed from the quote. He may have done that whether you asked or not, but reminding him that you're looking for it improves the odds of the price break showing up.

Getting the optimal labor and materials calculations takes the customer's help. If the job you're asking him to do is his bread-and-butter, you'll get a great price, especially if you telegraph that you know he is perfect for this job because it's . . . his bread and butter.

Future business

This is the best way to get a price break today. I'm dismayed how many times people leave this one on the table, meaning forget to use it.

All businesses value future work. Getting future business locked in means there is less need to manage level scheduling with huge price drops. The company can order materials more efficiently and get price breaks from suppliers. Having visibility to upcoming work also makes it easier to manage cash flow and thus reduce borrowing.

Case Study: After being elected Treasurer of a professional organization, at my first Board meeting I found one of the officers was negotiating with hotels for one meeting at a time, even though we met six times per year. She said it was in case we wanted to move around. We moved around, she explained, after hotels overcharged us.

That day, the hotel charged us $18.50 per urn for six urns of coffee for only twelve people over four hours (two sets of three urns). They had supplied coffee based on room size, not our attendance or consumption. Seeing that bill, I asked the Board if they liked his hotel. Everyone nodded. Off I went to the Hotel office with the bill in hand. By the time I was done we had six meetings per year, and only two urns of coffee for a $30 flat fee at all of them, retroactively including today. Between room rental and coffee, the bill was slashed in half.

Home improvement might not be such a rich source for future work unless you're a landlord or property manager, but there's some money there even for a single home. It doesn't need to be future work from you, just future work for them. If there's any chance that neighbors, family, anyone you know can be persuaded to throw business their way or even merely that you'll make a champion attempt to sway their decisions with your good words, let the company know at the estimating stage.

If this is a small starting project or you're getting a trial portion done, don't bury the lead. Tell him there's more in the future if we're happy with this order while still shaking hands at the first meeting. Let his hand go AFTER he has heard there is future work in the offing. Not only will this result in his best price, but it greatly improves the odds of having a smooth and hassle-free project.

Asking Questions

There's a commonly held misconception that asking the estimator a lot of 'dumb' questions will reveal ignorance and therefore peg one as a newbie that he can overcharge. People with this idea don't say much and tone down any inclination to look surprised when the estimator describes what he intends to do.

Asking all your questions is key to getting the lowest price. Don't worry about appearing unknowledgeable. It's too late anyway; he already knows. He knows in the first minute.

He already knows because every trade, installation or replacement has a major option choice or product quality level that the customer selects. When you don't say those things in the very first minute, he knows you're new at this.

When you say "I want" . . . a window, a deck, some landscaping, and stop right there, he knows. He can't quote without knowing if you want double hung or casement windows, if you want a wood or AZEK® deck, if you want sod or hydroseeding . . . you get the idea. He's got his work cut out for him explaining the options as well as the pros and cons of each.

Once he knows you are a blank slate, the more you stay silent, the worse it is. You think you're appearing knowledgeable by asking no questions, but from his point of view you're showing no judgment or screening. Blank check! He mentions an expensive optional process as if it's necessary, and when you don't raise an eyebrow, you just bought it.

For the newbie customer, the strongest negotiating position is to be a fast learner. By asking a lot of questions and spin-off questions from his answers, you become more formidable. You demonstrate you are able to speak up. It will start looking like pulling the wool over your eyes won't be easy. There you go. Back to a strong negotiating position.

Even better, when you phone the next shop for a quote, you know some stuff. You're not the person the first guy met.

With the second guy you mention your preferred option in the first minute. You say you're interested in the middle-quality choice. You want casement windows, you want a wood deck, you want sod, or whatever is the case. The odds of a too-high price have just decreased exponentially, because if he wants the work his pricing needs to be in the range a knowledgeable person would expect.

Need to Know –Err on the Side of Too Much

Some of the most stupid and costly mistakes happen when the homeowner shares the least possible site information with contractors and guesses wrong on what they need to know. Tell your contractor as much as you can about the history of the property and past work that has been done, even if you can't see how it involves this project. Also tell him what else is going on within a month or two of his project.

Often he can come up with something surprising to fit your needs or help out, something you would never guess. If you are going on vacation the day after his contract completion date, if you are having roofing done the same week as plumbing work, share that information.

Just because you 'can't see how it matters' doesn't mean squat. Let him tell YOU it won't matter to him. In the first case, knowing that the completion date is truly a drop dead date, he may order materials a week earlier to be extra certain late deliveries don't eat a day. In the second case, the plumber may need to work on the plumbing stack pipe on the roof, so must coordinate with the roofers.

If you tell him about pending projects, he might offer that his work should follow that other work rather than precede it. There are dozens of ways that giving your contractor the big picture can save you money and time. To have him say 'it won't matter' is hardly a huge downside compared to even a tiny chance that he can suggest something to improve the outcome of one or both projects.

Time and Materials

All products are time and materials. You will know the real cost of anything when you know the time (man-hours worked) and materials (purchased items).

Most of the time when you hire a contractor, it won't be for a huge project; it will be for fence repairs, a new deck, a new furnace, adding central air, painting, landscaping, bathroom remodeling, replacing a door, roofing, putting in a skylight, in other words all manner of home upkeep and improvements. Using the 80-20 rule, meaning focusing on what homeowners do 80% of the time, the next few chapters help you when the project involves one contractor or one-contractor-with-a-subcontractor. If you are considering a larger project taking weeks or months, be sure to read the *Major Renovations and Additions* chapter. While everything here also applies to more major work, when launching a big ticket expenditure there are additional steps.

A time and materials calculation[10] is the only way to be absolutely certain you are getting the best price. Every contract business uses this method. Absolutely every single one. If anyone tells you their company doesn't base prices on hours worked and cost of goods, it's because they're charging you double or triple that number and are terrified you'll do the math. Companies either bill you for time and materials or charge you much more. Or go bankrupt. Those are the three choices: too little, just right, too much. Only the middle choice is safe.

No one is going to go bankrupt by charging a 'full burden' time and materials. They will have a nice life, a good income, and grow the business. To land your job within a sixteen foot circle of the safest price, generate a

---

[10] This doesn't mean you buy without a stated price, just let them work on and on like Eldon the Painter did on Murphy Brown. You still obtain an estimate, and the contract you sign is for a stated dollar figure. When things outside the scope of the contract arise, either as change requests by you or huge 'discoveries' in-process, they are handled with signed change notices containing a price. Time and materials is how estimates are derived. Unless you are supervising constantly and counting only working minutes, docking for 'pondering' and the like, no open-ended pay-by-the-hour contracts.

time and materials for your job, make your contractor meet that price, and everyone eats well and pays their bills.

A full burden labor rate is not the hourly amount the workers make. All business expenses are covered in the full burden—insurance, vehicle payments, mortgage, utilities, supervisor pay, even profit margin. Nobody loses their shirt by being paid a full burden rate for their work. The full burden labor rate method enables a business to deep-dive into expenses and costs only once a year, and then for every job it's simply adding up the hours all the men work and multiplying by the labor rate. Small businesses hire an accountant to calculate their full burden rate annually. The full burden consists of expenses that will stay fairly fixed for the next year: mortgage, hourly pay of the worker, insurance—and those that vary but can be averaged: utilities, fuel, training, maintenance.

In this chapter, 'materials' will include equipment rentals (dumpsters, backhoes, etc.), disposal of scrap, safety supplies, disposables, and items used on this job. Things he wouldn't buy without this job. In a larger sense it includes permits too, although it's useful to list those separately, because they are so important.

Every dollar over time and materials either risk-factor money or opportunism.

Wanting to pay a price below the time and materials for the job is a price-wish. Everyone wants to pay $1000 for a new car and $100 for a 55" HDTV. But those are price wishes. Expressing a price wish does not have much impact on a seller. He can tell you simply don't know.

On the other hand, expecting a price too high will push it even higher. A contractor may test the waters by throwing out a price double what it should be, to gauge your reaction. If you don't react negatively, he can go higher yet.

A window replacement estimator might say a neighbor replaced all the windows in their house for $14,000. If you are not surprised or shocked at that price, he will land pretty close to that even though you're having nine windows replaced. If you start probing him with questions about that—"Is that at about three hours per window? Did they have any special or expensive windows? Did they all fit into the existing openings, or did you open up or lengthen any of the openings? Do you use foam or fiberglass insulation?"—He will know you're not going to let a too-high price slide by.

In addition to hours multiplied by the applicable labor rate, projects may have a risk factor[11] added in, permit costs, warranty and callback costs, and expenses unique to your project like equipment rental, dumpster rental and disposal costs.

---

[11] For a full list of risk factors, see *Addendum B, Ten Risk Factors* that increase the price.

Permit costs can be considerable. The permit cost roughly covers the cost of the Building Inspector's site time and record-keeping related to inspecting the job to ensure it is to code. Jobs like installing a circuit breaker panel or replacing a furnace require only one visit, but building an addition will require several visits at different stages. There is no price break if your house is five doors away from the BI's office, either. The charge is based on an average for the whole community.

To lean on the side of fairness, I add a $200 truck fee to each job. A week-long project may have two or three truck fees; it depends on complexity. It's a shorthand method to avoid deeply studying every trade. Doing a close time-and-materials is possible only by admitting there are some unknowns and throwing in a bit extra for those expenses. Taking the approach of assuming there are additional disposables, fasteners, procedures and whatnot is a safe bet.

Other additional costs include warranty and risk dollars. GC and contractor's warranties are generally of the 'self-insured' variety, meaning when issues arise during his warranty period, the contractor 'eats' the labor and materials costs necessary to fix it. Of course he hopes for perfection on each and every job, as do we all, but realistically that is impossible so he adds in a bit to cover callbacks.

It might seem logical that a perfect time and materials, with permit fee plus rentals, is the price to expect. You can try, but my experience is that the price with truck fee is the lowest practical price.

When the seller does the cost estimating, he doesn't use a truck fee; he works from charts and percentages and itemized lists. He knows his shrinkage and inventory carrying costs. He knows the warranty and callback annual averages. He has established averages for fasteners, disposable equipment and staging work, even weather. As a buyer on the other side of the fence we don't know any of that. But hitting the nail on the head isn't required, we just need to get close.

Getting close is a magnet that will have tremendous force on your estimator. If you have a price in mind that is too low, the estimator will dismiss it as a price wish and won't suspect you did some math. Too high, and he'll say "Sure!" and probably add another $300 on top to see if you balk, or pay it.

You aren't being ripped off if his price is $100 more than your careful time and materials calculation. An amount that size is within the day-to-day estimating variation and rounding numbers. You are trying to catch the $1,200 or even $8,000 overcharges. You don't really know if his full burden is $47 or $52 per hour, but it sure the heck isn't $210 per hour.

To get a feel for labor (time), TV shows on home improvements or flipping houses offer candid views of how long certain construction steps take. There are Youtube videos showing bite-size tasks like installing windows, replacing floors, putting up drywall, installing a skylight, and so on. Q&A websites and books are also good sources. Acquaintances who had similar work done will be glad to talk your ear off about it.

By one means or another every business expense must be covered between the labor and the material mark-up, because you, the customer, are the sole source of business income. Knowing this puts that businesses' luxurious landscaping, marble foyer and bank of shiny leased cars out front in a new light, doesn't it?

Time

The Full Burden Labor Rate includes:
   Worker's hourly rate
   Worker's paid vacation, training time and holiday pay
   Worker's benefits and insurance
   Supervisor's pay, insurance, benefits
   Sliver of the office staff wages & benefits
   Sliver of the mortgage, utilities, fuel, loans payments, taxes
   Sliver of top management's pay and perks
   Every business expense not materials for this job

A 20% markup, which includes unexpected one-time costs, cost of capital, reinvestment funds, and if any is left over, profit

Business expenses vary – do they rent, pay a mortgage, or own outright? Fancy office or not? On top of that comes owner's pay.

2013 baseline	Low skills	Medium skills	High skills	Electrician & Plumber	Designer	Architect
Wage, hour	8-12	10-17	14-40	25-45	30-60	50-150
Benefits, taxes, holiday	2-6	5-11	8-15	10-18	10-20	10-35
Supervision & support	1-6	6	9	12	3	10
Business expenses	8	8	11	8-11	8	12-40
Equipment, debt, overhead	5	5	7	5	5	5
General Mark-up	20%	20%	20%	20%	25%	25%
Full Burden hourly charge	30-44	41-57	59-100	72-110	70-120	110-300

Low Skills: housecleaners, laborers, trainees, using a shovel. Jobs not requiring significant math or writing skills.

Medium skills: jobs with longer on-the-job training. Jobs requiring measuring. Cement work, framing, HVAC, roofing, painting, installer.

High skills: jobs requiring off-worksite training, over ten years of experience, or high levels of responsibility. Artist, machinist, journeyman, Supervisor, bookkeeper, draftsman.

Use the low end of the range for your time calculations unless you have information on local averages or know what the business charges. The worker's hourly wage is only loosely related to the full burden.

In the spirit of being helpful, and agreeing up front that anyone who disputes any figure here is right, using these 2014 numbers will put you in the ballpark and that's the goal. These are very round numbers showing how full burden hourly charges might work out. It goes without saying that each business will vary in how many employees report to a Supervisor, how much equipment is needed, and so on.

Close study of this chart shows that for most of the manual semi-skilled trades, four times the hourly pay may be used as a rough full burden. In real life, a lower hourly charge doesn't necessarily mean a lower total cost. A talented or experienced person who earns more can get more done in fewer hours.

When you see how much time is spent in pondering and chatting on many jobs, you'll see how a few dollars more per hour can be worth it for a steady, experienced worker. Fewer hours are the inevitable result of more skill and less confusion about what to do next.

Something as simple as a leadman who calls out the next step while tying up the current one, instead of waiting until the workers are standing there looking blank can get eight minutes more work into an hour without anyone feeling like they are moving faster or rushed. When workers start in the morning and again after break, a clear direction or starting off with an easy task gets them moving again. Otherwise it can take up to fifteen minutes to get in the groove again, resulting in a half hour of lost labor per shift.

What if a shop offers a lower price than a fairly accurate time and materials? The only way this is legit is if it's their slow season and they are OK with break-even work to keep busy. Don't buy this reason without checking with other sources to make sure it is the slow season.

Otherwise you're dealing with a swindler or a business with worrisome business practices.

Worrisome business practices include:
  inadequate insurance
  no intention to pull a permit
  unskilled workers
  unpaid suppliers
  unlicensed in a trade that requires a license
  cash-only employees not covered by workman's comp
  skimping on equipment and safety
  misleading you about material quality

What if you get a super-low price because he's using a cutting-edge piece of equipment or a process that reduces labor? Time and materials still holds, it's just that instead of the usual twenty man-hours it's now three man-hours, or whatever the change entails. This has happened in manufacturing several times, but is rare in the construction trades. Do an online search for it, because it will be all over the news if there is some revolutionary product or process like that.

Jobs between 4 and 7 ½ hours

On-site jobs taking over four hours but less than eight may encounter the practice of charging for an eight hour day.

This sword cuts both ways, however. If the price makes you suspect your six and a half hour job is billed for eight hours, it's possible to add some work content for only the cost of materials.

Ask the estimator if your impression is correct that you're being billed for a full day, and if he looks a little uncomfortable, cut off his excuses and ask if you could throw in some more work content to fill eight hours.

Drywallers and painters typically cannot squeeze a second job into a spare two hours but could accomplish quite a bit more on the existing worksite. If you are paying for it, you might as well get some benefit.

Normally if you just ask for more tasks, the price goes up automatically, but if you indicate you are cognizant that eight hours of labor are already in the bill, now the two of you can discuss what else can be done in to fill out the day. If this is a job that takes three guys to finish in six and one half hours but is too much for two, there may be quite a lot that your three guys can fit into the time to make it stretch to eight hours.

This goes much smoother if requested before signing the contract instead of phoning at 1 PM on the day of work, but it could still work out well even at that stage.

Time and Materials

Overtime in Cost Estimating

Businesses that are prone to feast and famine, or peak times and no-work times often perform a significant portion of their workload with overtime. I'll share a cost estimator big secret: we don't care.

To us if the worker is a fully-burdened $41 per hour he's $41 per hour whether working straight time or time-and-a-half. The reason is that a company has a lot of behind the scenes costs related to this employee. These costs are covered in the first forty hours. The portion of the hourly rate that covers health insurance, vacation, workman's comp, utilities, rent, office staff, and loan payments is paid in full at the forty hour mark. Moving into time and a half means the employee gets more money, but the other charges aren't there. Social security and a few others go along with each hour worked, but still it's roughly even steven.

An estimator must add a bit to the quote if part of the job might be done on double-time. Double-time is usually given on Sundays and holidays. Some companies may have double-time kick in after a certain amount of overtime hours. When companies ask employees to forego vacation and work it they pay double-time or better. I've even seen triple-time when it was both Sunday and a holiday.

Sometimes a business assigns a higher rate for service performed outside of usual business hours; electricians, dentists and plumbers come to mind. They do it for another really good reason, and that's to discourage weekend, middle of the night and Sunday calls unless absolutely necessary. More power to them.

Easy Ways to Obtain Time

An estimator calculates the time and per-hour burden based on copious industry experience, grasp of local market conditions and intimate knowledge of his company's equipment status and debt burden. We're not going to do it that way.

Instead, we make two assumptions. They're really good assumptions. The first assumption is that his full burden hourly charge will be about on par with those of similar businesses in the region. If his is $10 more it will cause him to lose the bid too often, and if his is $10 less he'll leave money on the table all the time.

The second assumption is about the job duration. This tradesman's career involves variations on a theme within a niche of the construction industry –just cementwork, or just painting, or just landscaping, or just masonry, or just framing, to name a few. To him your job has a best way to go about it, and that way pretty much presents itself to everyone with a good grasp of the field. It may be a total mystery to you, but to him it's connect-the-dots. This means if you share the same information with each bidder and they are all good at it, their time estimates will be stunningly close. Within 5% is common.

So combine the two: if you know the full burden hourly charge of one company, it's good for all. If you find out the time duration from one experienced, honest vendor, that's how long it takes everyone to do that portion. Mix and match.

Ask the Estimator:

Obtain the time by asking an estimator process-related questions.

It is natural and acceptable to quiz the estimator on the actual work days and the process steps. It is also prudent to ask questions about the skill levels and experience of the workers. From this you can figure out if the worker(s) have low, medium or high skill levels and what hourly charge to use in your math.

Questions you might ask are below. Some will be appropriate for some jobs and not others. Don't pepper him with questions; pick out five or six that are most important to this job, and as conversation flows, a few more might get answered. When he figures out you are trying to gauge his time and materials he might clam up; however, using our second assumption above, pick up where you left off with the next estimator, and between them all you'll get a good view to the actual job. Questions:

How long will that take?

How much time will I need to take off work?

How much does XYZ cost? [a mantelpiece, a square of roof tiles, 100 ft. of Romex wire, a bay window, etc. Indulge your curiosity.]

How long does it take to remove the old one?

What happens to the old one? Or removed parts, leftover parts, etc.

What amount of days or half days will someone need to be present?

How long will that area will be out of commission?

How do they handle restroom breaks? [As an aside, many contractors plan on not troubling the homeowner, instead plan on having the workers drive to a gas station. Living two country miles and five stop signs from the nearest one, I'm not keen on having them disappear for twenty minutes to tinkle. What you do is up to you, but I insist they use my bathroom. Clear it of breakables and place a roll of paper towels on the counter for them to use.]

Do all your workers have on-the-job training only, or do they have more formal training?

Is the supervisor a journeyman?

How long has he been doing this?

What parts of the job will you subcontract?

How long have the helpers worked for you? [Finding out, for instance, that the work will be done with one skilled worker and the rest just assistant-level workers means there could be a $30 difference in their full burden hourly rates.]

What's the usual age of houses you work on?

Have you done any similar work in this neighborhood?

How will you get materials into the house? [Asking about who carries what and how the physical lifting will be performed not only helps you gauge manpower and equipment, it helps you spot potential problems: "You can't bring a truck back there, that's where my septic system is."]

Will there be any hot work? [welding, soldering, hot glue, open flames, etc.]

What safety precautions do you use?

What if you open the box and it's the wrong one? [this is asking about his return and delay-reducing procedures]

Do you do daily clean-up, or just at the end?

Between three or four estimators, each one revealing a portion of the whole, it's possible to get a very good feel for it.

Use a Third Party Source:

The second way to find out the time and materials costs is from someone who had similar work done. One source is the references the contractor supplied. Fish for others by talking about the work you are planning with co-workers, people at parties or any other venue to prompt people to share they know someone who did that recently.

A caveat on doing this: it's very rare to land upon a project that is close enough to yours to use the whole price as a yardstick. Usually there are major differences. Pick their brain about the process and collect any warnings they wish to provide. Knowing how their job differed helps you develop your price: is their house older or newer? Access easier or harder? Materials more or less expensive than the ones you want? More square footage or less?

Average per foot:

A third way to know a ballpark figure is by using per-foot averages. Construction has a per-square-foot, per-yard or per-pound average for certain projects that with a little hunting on-line can be found for your region of the country. For instance, for decks, paving stones, cement or asphalt work of all types there are square foot prices. Often they require a minimum charge, such as $900, to cover fixed costs.

Builders have a square-foot price for construction too. It might assume an average level of quality for the region which could be skewed by the maple flooring, brick to-the-cathedral-ceiling fireplaces and half-circle windows of upscale homes, so if yours is a family room addition with ordinary double-hung windows, no bathroom and no fireplace, expect your per-foot price to be much less. Be aware that the kitchen cabinets and countertop can be 15% of the new-house cost, so pooling those costs into a non-kitchen addition overprices the thing right off the bat.

All trade estimates by square or linear foot have low quantity and high quantity prices and the price is lower the larger the quantity. There is no case where you pay a premium because the job is large.

For unit pricing, the price should ALWAYS drop when the quantity goes up. It doesn't matter what the thing is: bricks, trees, paneling, shingles, hardwood, shrubs, paint, tree removals, windows. The fellow with the 200 foot long driveway may pay more to have it paved than the guy with the fifty foot driveway, but the per-foot price should be lower. This is true of service work too: duct cleaning, window washing, driveway plowing, cleaning lady visits or termite treatments to name a few. If two neighbors join up to have their driveways paved with the same RFQ, the cost will be lower for each than if quoted separately. Negotiate a year's worth of service or combine several neighbors into one service call for big savings.

Industry average:

Some jobs are too complex to develop a good time and materials. Jobs requiring coordination of several subcontractors, such as building a house or a large home addition, can be benchmarked against similar jobs or against the industry average published for the state or county. Sometimes these are grouped as high end or low end of the market.

It may be useful to estimate time and materials where you can, perhaps 70% of the job, and fudge factor the remainder. The purpose is to catch the huge out of bounds figures, 30% to 100% too high.

Industry averages that include data from before 2008 are skewed to an unreasonable maximum.

Construction has, in the past, been based upon what the market will bear. During the housing boom several years back it was common for a builder to pull a $300,000 payday from a $700,000 house. That kind of mark-up might seem criminal, but without it new construction was far cheaper than buying a 25 year old house because the market was so out of whack. Even with today's lower home prices a builder can net a healthy six figure income building three to five houses per year. When using industry averages for your area, be sure the neighborhoods and date ranges will render a useful number for your neighborhood and situation.

Reverse estimating:

The fifth way is to start with the quote amount, then subtract the retail cost of materials. If relevant, take off a truck fee or a couple of them. What is left is ostensibly labor. Divide this by the hourly rate of your estimation of the skill level and see what hours you get. Ask yourself, does that sound right? If even at this point you have no idea, you can call the guy and ask "Someone said I will have workers crawling around my yard for six weeks. Is that true?" He might reassure you, "Oh no, we'll have it done in two weeks, my guys will work Saturdays too until it's done." Now you have an idea. You can check even further with "It's OK to have two or so guys using my bathroom, but if it's ten guys, well it would be good to know now." And again he mollifies you with "Oh no, it will just be two guys."

If you place your area of concern slightly to the side, you can get a clear picture of the labor. Ask about your days off work, will they be around on weekends, is there going to be any ladder usage and who will hold it? Can I provide cold water bottles to the guys, how many will there be? (be sure you give them the bottles if you told the boss you would)

Plumbers, electricians and handymen will tell you the time involved within fifteen minutes, practically. They don't want the back and forth, and they just put their profit margin and parts mark-up in the hourly rate and that's it. They lay it all out there, $110 an hour and it will take 45 minutes plus an $80 show-up fee. They don't really make $110 an hour because they spend eight hours a week replenishing supplies, doing billing and accounts payable, plus seven hours talking on the phone or providing estimates for nary a dime. They often take no mark-up on parts because it annoys customers who upon seeing the bill wish they had done the shopping themselves.

Case Study

The job entailed installing a stone front, new mantle and hearth on my living room fireplace. The estimator walked through the process for me, and said on the second day they should be all done and cleaned up by 11 AM. After I asked about the weight and heavy lifting, he said two guys would handle it quite well. A few days later he sent a quote for $2,200.

Another bidder had shared that he would charge about $450 for the required boxes of stones needed which included an extra two or three boxes " . . because you need to get the right look when fitting the stones, size, color, shape all need to work together. It's like an art." That made a lot of sense, so later on it wasn't troubling when there were about 20 leftover stones.

My selected mantle was $180 retail, the hearthstone about the same, then wire mesh and framing. Assuming he gets a price break from retail is reasonable.

After deciding to give him the job, I told him their huge portfolio of jobs similar to mine was impressive so I would love to give them the job, but two guys at $40 an hour for twelve hours plus $800 in materials and a $250 truck fee comes to $2000. Then I stopped talking. After about fifteen seconds he said "Sure." We had a deal. A photo of my beautiful new fireplace went into his portfolio too.

Could I have gone for $1800, taking into account the probable materials price break and only a $150 truck fee? Perhaps. If the goal is the lowest possible price, then go for it. Having been around the block many times, my goal is a fair price that keeps the guy in business until next year, in case there is warranty work or another job. Sometimes an owner just isn't able to stand tall and turn down a job when the margin disappears.

Besides, on reflection, one good reason for a more modest price cut is because I don't want to start a back and forth. I just want him to say OK. Spelling out the math is the way to reduce objections. If the initial counteroffer is too low he will be compelled to start a back and forth. It's not about being conflict adverse; it's about paying fairly for a man's labor.

In the case above, there is no possible way I was ripped off; I witnessed the labor, and there was mortar and tools and other things not itemized in the materials cost, which was obtained from a third party, not him. A huge part of their value was that their skill set matched my vision of the outcome. Was that skill set worth the extra $200 out of respect? Sure. But I'm not made of money.

When they're double or triple the good price, time and materials will help you spot it immediately. When they could come down a bit, use time and materials to get a lower price by sharing the numbers with them. Don't just counter with a lower total price, walk them through in a very rough manner why they could get by just fine with a lower price. I've done this many times, and the stories I'm not telling you are the thirty times their estimator countered with 'yes, but you didn't include the $400 we pay to the electrician' or 'I don't know where you got your slurry twin pricing from, but we use Mach II slurry twin that costs twenty-two dollars a gallon' or 'I can only do ninety pieces per carbide insert, three inserts at a time and they cost seventeen dollars each, so I work that into the per-piece price.'

By sharing the numbers with them it's more of a discussion about process than a bicker about total price. Invariably what he blurts out can be confirmed by an internet search or phone call, and it's so off-the-cuff and unrehearsed that it's bound to be true. Generally my response is 'oh.' If the guy is about my age and we have some rapport I'll do my Gilda Radner/Emily Latella imitation of 'Never mind.'

Maybe my prickly toolmaker background makes me more tolerant of the gruff owner who gets his back up when I missed the electrician cost or used a too-low materials cost, but I tend to like him better when he blurts out his real cost without artifice.

Materials

Look them up on the internet. Search for similar jobs in your region of the country, hardware store pricing, or supply house pricing. One useful source of detailed specs and pricing is town and city government websites. They will often have their repair RFQ and contracts on-line, with full disclosure of pricing. Those contracts will reflect better pricing than most homeowners can get because they have large quantities and repeat business going for them, but getting close to their deal is a win.

Government entities do all kinds of work to buildings and grounds that parallel homeowner projects. Another plus is if you see a name attached to the project, calling the Government employee during work hours with questions such as how happy were they with that contractor stands a good chance of getting helpful answers, since responding to questions about the project is part of their job.

The first time you do a time and materials, you may think you've done it wrong. After getting multiple quotes, lowest quote is for $20,000. The work involves three guys for a week, actually two guys except for the first day, and one of them is an assistant making minimum wage. The materials cost roughly $4,200.

So being generous, 3 guys x 40 hour week = 120 hours at say $70 an hour average (the assistant less, the leadman more), that's $8,400 plus the $4,200 materials, for a total of $12,600. Would all of them really overcharge that much? This must be a mistake!

Nope, no mistake, you get why this works now. Add in a truck fee, another $600 for permits and disposables, but you're still looking at $15,000 being a very generous price. Nobody is going to lose their shirt or not make payroll this month at $15,000 for this job. That's why you can go back to him and say, "I'd love to hire you, you are a top firm for this job, but using three guys, 110 hours tops, with $70 an hour being probably too generous, plus $4,500 retail for the hardware and supplies, all this makes $14,000 more in the ballpark. What do you think?"

He might come out with something surprising like "Oh, I'm not using the El Cheapo, I'm using the El Grande that retails at $6,500." So you could say, "Uhh, I'd rather have El Cheapo." Or you could say, "OK, so that puts us up at (calculator clicking) more like $16,000. Thinking about it, the job won't need all three guys for the whole week, some days it will only be two, right? So, it looks like a ballpark price would be $15,500 for the El Grande."

It won't be a who-blinks-first. It won't be him talking you around in circles. Cold hard numbers. He's got to produce cold hard numbers to support his calcs. If you've picked the best guy for the job, tell him he's the guy you want to work with, you feel sure he can do a good job. Tickling him with a feather while you make him take a fair price really works. Figuratively, I mean.

What if he says 'my full burden is $100 an hour'—you simply reply 'that isn't industry standard for this field.'

It's best if the high price never hits the light of day by letting him know there are multiple quotes and you're cognizant of the time and materials. If he tries it anyway, the time and materials approach makes it a very brief talk because you're on firm footing.

Job Duration Tips

The hours of labor do not add up to the duration. Nearly always, there is some idle time and some air time. Idle time happens when workers are stalled to a stop when they could be working. Maybe they ran out of materials, a tool broke, the customer requested a stop or the proper equipment is missing so now they stand around.

Non-working time caused by workers taking long breaks, shooting the breeze, or leaving the worksite for no work-related reason is called screwing off.

Air time is when the job is not progressing because no one is working on this job during normal working hours. No need to demand a reason from your contractor, I'll tell you what causes is right now: poor planning. He will never tell you that. He will make up something. You will twist yourself into a pretzel trying to follow the trail of his bogus reasons. Even if there is another reason besides poor planning, treating that as the reason puts your head in the right place to get your project progressing again.

Poor planning means the crews are scheduled like it would never rain or the truck would never get a flat tire, so a three-hour delay snowballs into affecting six customers. It could be lack of thinking ahead on the resources needed for this stage of the project, or materials that aren't delivered on time. It could be extra time wasn't allotted to less-experienced workers; it could be no slack was left in the schedule for callbacks, so in-process jobs incur delay after delay as those callbacks are handled.

Additional trips to worksites balloon into huge amounts of lost hours until managers lose hope of having satisfied customers so stop trying. This explains the ire that greets the customer who simply asks for some workers to show up to finish the job. If you have to ask, it's already clear your project is being mismanaged. The squeaky wheel will get the grease.

Projects are like making a big holiday dinner; some things are started the night before, some are started at 10 AM, and some are started at 5 PM so they all come together at 5:30 PM. It's not possible to be a good cook without being a good planner. If you start boiling the carrots three hours before dinner but put the duck in the oven one hour before dinner, you'll have neither palatable carrots or duck. The long leadtime tasks must be started earlier, and short leadtime tasks later.

Poor timing is the primary reason for two month long kitchen remodels. Sure, vendors lie about their delivery times, but that's where experience comes in. An experienced contractor learns which vendors are likely to be fibbing when they say 'it'll be there in three weeks" and plan for five weeks. There is no such thing as a two month kitchen remodel; there is just a ten day kitchen remodel with a lot of air time, sloppy material ordering, inadequate manpower, slipshod measuring, and wasted effort between the start and finish.

GC often do not see themselves as serving the needs of the workers, to keep them moving steady and stress-free, but as serving other masters: gaining quantity breaks on materials, tax shenanigans, or quarterly earnings timing. If you are dealing with a contractor focused on shaving 50¢ off the box price of this or that while his workers onsite are standing idle for half an hour waiting for supplies, he's spending his time with a faucet drip while the outside hose is open full blast. If he has what they need available when they need it, he will be far more profitable than if he saves $50 on lumber by buying from a place 30 miles farther away.

A customer should never be charged for air time but might be racking up charges for idle time, depending upon the cause. A contractor who has greatly overcharged you will almost always throw air time into the job so the start and finish are further apart and actual labor hours are not obvious. He's making it look like you're getting your money's worth when all you're getting is worse service.

Some tasks have cure, settling, or drying times, so a quiet day is appropriate. After painting, it's best to avoid movement that stirs up dust. Most types of flooring need a day to set. Workers may either work on some other part, or leave the premises to work elsewhere. If there's a point when another trade or subcontractor, such as an electrician, needs to come in, a contractor might leave a large gap in the schedule just in case he isn't timely. He doesn't want a subcontractor's lateness making him look bad.

A shop might tell you the area will be out of commission for a longer duration than necessary because things happen; workers get the flu, equipment breaks down, supplies don't arrive on time, trucks get a flat. If all goes well the job is finished ahead of schedule, but if not, it still completes on time.

If days of duration are more than twice the working hours, the fellow is nesting too many concurrent jobs, is not coordinating well with his subcontractors, or is not ordering materials efficiently. Complain. Often.

The only way this gets better is to make daily or twice-daily reminder calls until your job progresses. Silently fuming over the delays is pointless; every day, phone him. It can be to remind him you expect service or just to say a cheerful "Any news on our delivery?" He's going to say "Just like I told you yesterday . . . " but too bad. He might roll his eyes when he sees your number on his phone, but this works. His future customers thank you because he will improve on future jobs so this doesn't happen again.

The Request for Quote—RFQ

The RFQ is a time-saving document that anyone can use to get quotes. It's the best choice for more complex jobs with safety and building code aspects.

Obtaining multiple quotes is less effort than you imagine. The first one is always 40-60% more work than the ensuing ones. After 20 years and hundreds of projects it's still true for me. It is an immutable law no matter how many projects one does. By the fifth quote it will feel like very little trouble to add a sixth. Take me up on this, you'll see for yourself.

The less you know, the more quotes you need. To get five good quotes, send out nine to fifteen RFQ. Some will be to the shops that don't specialize in your kind of project or don't want it for one reason or another.

How many RFQ to send out in your particular case depends on how many competing businesses are in range and the odds of getting a reply.

I'm not going to get competitive quotes. If that's your attitude, at the very least do not say or do anything to reveal that to your estimator. Opportunism will kick in. Besides, in 95% of home repairs each additional quote takes an hour. All the big thinking is done during the first one; after that it's just a shopping list. A project of any size might easily save $400 with more bids—even if it's to get that first guy to match the price. 1 hour = more than $200 most of the time.

An RFQ[12] is an efficient means to locate the few shops for whom this work is right up their alley while wasting only about one minute of the inappropriate shop's time. It is less wearing on you than calling everyone, playing phone tag and hoping you remembered to relay everything. Because all the facts and numbers are on the page, follow-up phone calls are much easier.

The aim of the RFQ is to provide each estimator with the same information, request the same scope and materials, address concerns that could add cost and make it clear which unskilled portions of the job you will

---

[12] Two examples of RFQ are located in Addendum A. Yours can be less or more detailed. One of the primary objectives of the written RFQ is conveying there are multiple bidders, immediately reducing opportunism.

do yourself, if that's your plan. One can even ask for off-season or non-peak price breaks without it feeling awkward.

If you phone, the wrong-for-this-job business will provide a quote, even if he wouldn't have with an email. When you phone, it's possible none are a good shop for this work; you'll pay a highball price even after getting multiple quotes!

What if you have to accept the quote from a highballer? It's more money but the work will get done because he priced it so he can partially or entirely outsource it. Put another way, when giving the job to a highballer, you pay a premium for divesting the responsibility of finding the correct contractor. It means the highballer firm is a middleman. Don't be surprised, however, if even after agreeing to pay his price he still doesn't want it!

Before signing a contract, make the RFQ an Addendum to the contract. Name it something, then reference it, with a note penned in the margin if needed, and both parties initial and date that note. If the RFQ is not a mentioned and initialed part of the signed contract, the specs and requirements mentioned in it are not part of the agreement in a legal sense. This means you cannot hold his feet to the fire later for failing to provide something requested in the RFQ.

Whether contacted by phone or RFQ, and whether there is a site visit or not, the shop should respond with a written quote on their letterhead stationery, mailed, faxed or emailed. Ask for a hard copy if he gives you an estimate over the phone.

Examples of projects better served with an RFQ are home additions, landscaping, kitchen remodeling, improving drainage, cement and asphalt work, adding windows or doors, and any project containing requirements about safety, work hours, material quality and treatment of the existing structure or best practices for foundations and underpinnings.

Small businesses, landlords and condos are better served by issuing RFQ for almost every project since a paper trail is useful for tax purposes as well as general good record-keeping.

Even if jobs can be set up or handled with phone calls, not an RFQ, it's very useful to make bullet points on a piece of paper, write down measurements or models to easily reference during the call, and then write down any good ideas you pick up along the way.

Content of the RFQ

>Your address and full contact information & date you sent it
>
>Vendor name, address, phone
>
>Words Request for Quote centered, in 16 or 18 pt. text
>
>Due date for the response and what form you wish it (email, letter, fax)
>
>Dealbreakers, must-haves, rock-solid deadlines (if any)
>
>Dates: start, end, milestones if known, dates to avoid
>
>Description of the work, any brand names and models involved, sizes, measurements (or rough square footage), specs, tests, packaging, things not allowed, minimums, maximums. Reference attached documents (prints, photos, documents)
>
>Mention if the AIA contract will be used (see the Contracts section)
>
>Equipment on site he may use
>
>Any special circumstances that may affect scheduling, access, or restrict employees
>
>Options to be quoted separately
>
>Specifics that you want his response to include.
>
>Request for references or examples of previous work (if applicable)
>
>Sincerely,
>
>Your Name

Due Date for the Response: This should be several days from the date he receives it. If he needs to take a look at the situation prior to developing the quote, include some mention of your availability: "Available to show the area weekdays after 3 PM and anytime Sat., call to arrange." Often a time like 7 AM or 8 AM works well too.

To shorten lead time, phone several shops to check if they can meet the timeframe, and if yes, set up site visits, then email the RFQ after the call or hand it to them when they arrive. An estimator may need to get current pricing on materials before he quotes, so allowing one to eight business days after his site visit is reasonable.

Be prepared to receive most of the responses right on the due date. It doesn't hurt to phone each one about three days before the due date to ask if he is considering quoting. If he tells you a quote will be coming, you might ask him for his thoughts. If it results in new information that affects what you want, send a follow-up email or fax to all the bidders right away. Quotes that are for different things slow down the process and are not comparable.

Dealbreakers: Dealbreakers are those things you really need. Perhaps you are looking for a shop that does two things well, will not outsource one of them. If the installation is tricky or the skill is more akin to an artist than paint by numbers, providing three references from recent similar installations so you don't suffer beginner's mistakes could be mandatory.

Dates are often dealbreakers. Put dealbreakers front and center, don't bury them here and there in the body of the RFQ. When composing the RFQ on the computer, copy the dealbreakers to the beginning and Bold that text. It's OK to repeat them later in the body of the RFQ.

Dates: In many cases the contractor is asked to provide lead times and dates, and part of the decision is based on the desirability of their dates. This section might say: Provide estimated best start date and days to completion after start.

Timeframe may be expressed as days after signing of contract, such as "start to be within fifteen days of contract signing." Beware a start date expressed as 'after payment of a 25% deposit', or 'immediately' when materials have a long lead time. There is no benefit to you to live with a teardown situation while waiting for materials. The contractor does this just to have you over a barrel until materials come in. Unfortunately, most kitchen remodels are done this way. The actual physical work, proceeding in a workmanlike fashion, takes two-three guys a handful of eight-hour days, but who gets the thing done in under two months? They dive right into teardown even though the materials have a six week leadtime so you can't cancel and hire someone else.

Description[13] of work: This is all the information and preferences you would like to mention. Be generous with words. Get photos if you want it to look like something you saw.

Don't make this one huge paragraph. Break it up. A half-page of 10 point text from margin to margin almost guarantees something significant will be missed. Don't worry about perfect groupings. Make a new paragraph every three or four sentences. Put an empty line between each paragraph.

If specs and details are more than half a page, consider making an attachment and reference it here.

AIA contract: This is mainly for larger additions or projects that take over two weeks and involve some design or a hierarchy of subcontractors. Let the contractor know that you will be the writer of this contract, not him. He might balk at this. Be very suspicious if he does. He might, however, use the AIA contract himself.

Equipment he may use: This is a good-guy thing to do. If you have a long ladder, a gas generator or a fifty foot heavy duty extension cord, may as

---

[13] Examples of RFQ are located in Addendum A – Forms

well mention it. On business locations it might be forklifts and cranes, drill presses to put in an extra hole on the spot, 240V power. Workmen use their tools a lot and suffer breakage. When possible, rather than incur a minor work stoppage it's worthwhile to offer your own drill bits, taps, saws, cordless drill, tape measure, cable ties, box cutter, wire stripper, Ziploc baggie, vacuum cleaner, screw driver or stepstool without hesitation. You don't need to mention these, but if the chance arises to save fifteen minutes by offering up yours, the job is done quicker and your workers are less stressed out.

Any special circumstances that may affect scheduling, access, or restrict employees. This will improve contractor efficiency, even though it will take a bit of thought on your part. Mention situations to be worked around, such as the week-long festival that barricades Main St. Perhaps there is a date range during which the job either can be fully completed before or can be started after, but cannot be partially done during that time, such as during an Open House or the family reunion weekend. It doesn't hurt to mention this not only in the dealbreaker section but here too. Mentioning it to the crew when they arrive in case it didn't filter down to them is also a good idea. Most projects won't have special circumstances.

Options to be quoted separately. This can be an extra you haven't made up your mind about or will use the quote to decide which way to go based on cost. For example, when getting a new kitchen counter, a person might ask for pricing a composite sink and then how much less for a metal undermount sink.

In the area of home improvements, a contractor may wish to know the lay of the land, figuratively, before pricing options. Without being coy about it, how much an option costs can depend upon how much one of you wants it to cost. I'm not saying the contractor is doing anything wrong. I'm saying if during a kitchen remodel one half of a couple says we can have marble countertops only if it costs under $500 more than Corian®, the contractor wink-wink can catch a great sale and do it for $499 over Corian. The total project will encompass materials costs without it being too noticeable.

A lot of options get done this way if a customer lets the contractor know the tipping point. There's no downside to the contractor because if the buyer declines the option in the end, the extra material cost is still in the main quote. He can similarly make an option very unattractive with his pricing. Even if there aren't two buyers with differing ideas, it can be immensely satisfying to a person to get a great deal on a desired upgrade. If a contractor can provide that little joy, why not.

Basically, pricing on options is influenced by what the customer considers an acceptable price, what the contractor prefers installing, and trailing behind in third place is the actual time and materials difference.

Specifics on what you want his response to include. If you wish certain elements to be batched in the pricing, or want to know how many workers will be on site or any other information he should include, ask for it here. For example you can write: Please provide 1) three possible start dates, 2) duration of work, 3) pricing for the base job 4) pricing on the three options listed 5) list the kind of material samples he has at the office. This helps the estimator draft the response when the RFQ is a few pages long.

Request for references or examples of previous work (if applicable). Omit this part unless you really will call or visit each one. If you ask for three examples and sent out seven RFQ, that would amount to twenty-one calls or visits. That's a lot of driving around. A note saying "The preferred contractor will be asked to provide three references involving similar work" gives them a heads up and more time to choose a few.

Sending the RFQ

A 40% response rate is typical. Remember that pressing unwilling vendors to quote will result in a quote—an untrustworthy, half-assed quote. Experience has proven pestering for a quote is a kettle of worms.

If the business isn't interested in working for me upon hearing the job particulars, it's very unlikely that a begged-for quote will rise to the top. It's OK to call a few days after sending it to make sure they received the RFQ, but just have the phone answering person verify it in fifteen seconds, don't make it an "I want to speak to the President" call. If they suddenly want to talk to you about it, great, have the conversation. Just remember if any improvements are developed on the spot, forward those new specs to all the other bidders.

Aim for five good quotes minimum when you are a first timer or out of your element; with three it is difficult to make heads or tails of price variations. With five, patterns appear. Having five or more sheds light on what is affecting the pricing differences.

Highballers and swindlers don't count as one of the five. If business is good, companies will be cherry-picking customers. It's important to avoid inexperience, and at this stage you can't tell if a company is highballing because they don't want the job or packed the quote with learning curve money because have never done this kind of work before.

RFQ Follow-Up

With those who are game to quote, have a phone discussion or site visit, then eliminate some of the bidders. Send the two or three remaining bidders the modified RFQ with more detail based on all the improvements you've collected from talking to bidders and getting more information from the internet, library, or co-workers.

Ask which parts of the job will be subcontracted.

Review the bids to ensure they include every task. Try to find any gaps between their proposal and your RFQ.

A choice between two competent candidates is a very strong position for a homeowner. At this point you could tell each it's a fifty-fifty situation, that others have fallen out of contention. Let them know that it's worth their time to sharpen their pencil.

When you remove a few from consideration, it's appropriate to thank those who submitted quotes but did not make the cut. It can be a kindly one-sentence email or voicemail message.

Contact each of the finalists to find out their approaches, timeframes and other details that can improve your satisfaction with the process and the outcome. This is your own personal 'Survivor' contest, meaning some good guys won't make the cut for almost no reason at all. When it is narrowed down to two contestants, pick one.

Quotes usually have a 60 day lifespan but in practice many companies honor them much longer if asked nicely and raw material prices have remained stable.

An estimator cannot really know how quickly the customer will decide or when the job will start, so a price break for positioning it in the slow time can't be offered. If you want this break you'll need to have that discussion with the estimator and let him choose the start date and duration.

It might occur to you to write on the RFQ, "quotes not containing pricing broken into these components will be discarded." Unfortunately, enforcing that could rule out the best contractors. Despite a clear RFQ and simple instructions, don't be surprised if not a single one complies with your pricing and details requests. After getting the quote, a line by line discussion of the RFQ requirements is in order. Finding out the parts that are not included in this quote is the first step in getting all your bids apples-to apples-instead of apples-to-oranges.

RFQ Obedience

At this point a person might ask, why bother? Why not just rule out the ones who don't follow directions and give the job to only those that do? You may have put thought into the RFQ, with dimensions and materials with photos attached and his quote comes back as a dollar figure and one line "Remove old and install new deck." This is common. If you want more depth, you'll have to talk to him.

But that doesn't answer the question. The main reason for not handing the job to the one who follows RFQ instructions and ruling out all the disobedient ones is because . . . experience has proven RFQ obedience is unrelated to quality of work.

Mr. Remove-old-and-install-new-deck may do an excellent job. He may comply with every one of your job requirements. He just doesn't see the need to repeat all your details back again in his own words, especially when his typing skills are hunt and peck.

RFQ obedience and high-quality manual work are really two different skill sets. There may be little correlation between misspellings on paper and quality of boots-on-ground work.

A good analogy is a barber with English as a second language. He could be a great barber but speak heavily accented English. No one makes the mistake of judging a hair stylist or barber by the size of his vocabulary and the perfection of his pronunciation. Even less by how well he types! There are some snobby factions who prefer heavy European accents in their hair stylists.

Most of us will agree hearing a hair stylist string English grammar together incorrectly tells us nothing about his skill at his trade. He may have won contests and earned awards for his coiffure work. The same logic holds for most tradesmen who work with their hands. It isn't that poor speaking or writing skills make him very good, or bad, it's merely that it's irrelevant.

The issue may not even be disobedience. If you have received six quotes and none of them included every task in the RFQ, hold off on calling each company to berate them for sloppy quoting.

When it has tasks from too many trades, turning the recipient into a General Contractor, he often simply quote his part. A landscaper is not going to bid on painting the house and fixing the roof even if it's in the RFQ. He might have no problem painting the decorative bridge in the rock garden or hiring a blacksmith to repair an old iron gate. If some bids no-quote a portion, ask the other shops that include it if they're doing it in-house or subcontracting.

Give hard thought to securing the missing or always-subcontracted part yourself. Before dismissing this idea, dedicate a modest amount of time, such as two hours, to learn about this industry or field. It is very likely that by the end of that time you will understand what to look for and what to avoid in that field and will be able to handle selecting that vendor yourself. Find their professional organization website with customer ratings. Co-workers or relatives may have good recommendations.

A contractor omits some of the work from his quote because he doesn't have a standard guy for that task. He doesn't have to say that out loud; if he always threw this portion of the work to his friend he wouldn't be omitting the task from his quote.

You can save money by sorting out each trade and RFQing each one. That said, the reason for not doing this is coordination and clear lines of responsibility. Some types of projects pair two trades, and in those trades they get to know each other and have preferences. If you hire one, the lines of responsibility are clear. You're paying for an outcome and he makes that happen. If you hire both independently and introduce them, asking them either to coordinate their efforts or to vet their schedules through you so you can coordinate them, maybe it's OK or maybe they hate each other. If things go awry the odds are high there will be finger-pointing at each other as the cause. The owner playing Solomon will not mollify the loser or be that compelling.

So, without contradicting myself even if it sounds like it, when tasks go together as a pair, hire one and allow him to subcontract the other. They are likely to have matching quality levels and work well together. It will be seamless to you. Still, obtain a lien release from the subcontractor before final payment for protection against for payment issues between them.

One more word about hiring the different tradesmen yourself: when you do, it's your job to proactively set schedules so they don't get under each other's feet. For instance, one can't be turning off the power while the other needs to use power tools. If you can give one of the guys his own day or row of days to do his work, so much the better.

Giving a tradesman his own day for work may involve timing and ensuring other tradesmen finish their part on schedule. One reminder isn't going to cut it. If things are delayed, your jump-in tradesman is going to need more notification than a call the evening before his work date. Once it's too late to secure other work, he will often charge you for that day and then charge again for the actual work.

RFQ Special Situations

Design RFQ What if the reason for needing a professional is because their expertise on possibilities is needed? What if you want some good ideas on how to improve and do not have specifications? For example, say you have a big flat back yard and you want something visually interesting back there. This is a perfect use for an RFQ instead of a phone call.

One of the goals of the wanting-ideas RFQ is to clearly show the level of service needed to complete the project. You want those who don't fit the bill to bow out while attracting those who have skills in the majority of your requirements. You want them to demonstrate they have preferred subcontractors for the remainder. The fellow who has to outsource too much or scoot up the learning curve on your job can be spotted in a side-by-side, apples-to-apples comparison.

In the big flat back yard scenario, the goal is to attract land-scapers whose design creativity meshes with your tastes. The RFQ would include the size of the area and a ballpark spending amount. Perhaps you're willing to have something taking up the whole half acre, but would also be OK with leaving the back as grass and concentrating the visually interesting part in the sixty feet nearest the house. What you would like is for them to pop over for a look at any time and then receive sketches, computer renditions, or photographs of some of their past work similar in scale and price to yours.

What you can't expect is the full design without pay. Truly, if they can provide three to five photos showing ideas they would incorporate, plus they can state the names of vendors who carry the key materials, they are on the short list. Knowing the vendors is critical; showing you a lovely trestle or aviary but being unable to find out where to buy it, or buy the parts, makes them worse than useless; you've wasted time on them.

It's frightfully easy to scam you by showing you a magazine photo of a greenhouse, tell you it will cost only $15,000, but in the end the contractor cannot find it. If he can't find it now, he never really knew the price. He was just getting your head in the noose by telling you something you wanted to hear.

Without an actual company name and address, or better yet a brochure, never believe they will 'find it on the internet' unless they actually do it right now. It's common to embark on a hunt for an item viewed in a magazine and dead-end with 'it's a stock photo" and "I'm sorry, we can't help you," Make sure your chosen shop has all their ducks in a row to give you what is promised. No matter how sorry they are upon finding out the company that built the gazebo you love is defunct, it's hard to forgive because they could have found that out with one phone call prior to signing a contract. The disappointment is not lessened.

But back to developing the concept RFQ. As you draft the RFQ you'll find you have more do's and don'ts than you originally thought. You may want to keep the three year old deck and want to have an attractive grill area. Plus, you want at least one little covered or shady area. It could be a gazebo or a tiki bar, you have no strong feelings about it yet. Then you include that you want an area for annual flowers, but you want most of the rest to be perennials. You add information on what you don't like, such as white stone mulch or those mini picket fence trims around mulch areas. Maybe you find a humpy berm surrounding three tipped white birch trees to be the height of tacky. Or you must have one.

Then you think about it some more and decide it would be wonderful if something bushy was located where it blocked the neighbor's view into your bathroom window. It becomes a requirement. You add that it doesn't need to be nine feet high right away, but it should grow that tall within three years. Also, it should do its duty all year around, not be a couple of twigs five months of the year.

Revealing your budget number or project maximum figure is a terrible idea with most contractors, but there are exceptions. A blank slate landscaping project is one. Landscaping projects fitting the description could range from $5,000 to $400,000. He could build a Thai temple of a barbeque pit. It's wasting everyone's time to make him guess how much you have to spend.

If you think the most you can afford is $20,000, start with $17,000. That provides a little room to upgrade on the size of shrubs or improve some materials. This project will be judged on how much magic each firm can wring from the same amount of money. Seeing examples of their previous work will be vital. Their talent meshing with your personal preference is the brass ring.

Some of them will call you to bounce ideas around or ask if you're OK with a September start. Be honest with what will make you happy; don't string along a guy unlikely to make the cut. You can give the job to only one, so don't hesitate to narrow the field and save some guy a bit of fruitless work. If you want to enjoy it in August, a September start is out of the question.

At that point you thank him for his honesty and tell him you regret it won't work this time, but you will certainly keep him in mind for future projects. Contractors who tell you they can't meet your dealbreakers are not bad guys, they are honest guys. Relatively more honest than they could have been.

Sometimes a bidder will press you to decide between two or three things prior to jogging up the quote. It's OK to ask him to jog it up the way he prefers to build it; if that choice doesn't make the cut, you can ask him to modify it during the phase when you take all your apples-to-oranges quotes and make them apples-to-apples.

Reviewing Bids, picking one

Remember contractor heaven: if your preferred vendor isn't the lowest bid, don't give up on him without giving him a call. Give him a chance to get the work.

One possible way to say it is, "I am impressed with your knowledge (or whatever is true, the reason you want him to be a viable candidate), but this quote is a little on the high side. If you could sharpen your pencil, I'd give you the job."

Ethics

There are no ethics issues surrounding going after a plain and simple lower price. Ethics issues are related to overcharging or proposing under-the-table deals. Stating that you can get a lower price elsewhere is neither an ethics issue nor a big secret. As a buyer you are allowed to buy from who you choose for whatever reason suits your fancy.

That said, as best you can, wear the white hat. Do not pick fights with contractors, especially during the estimator visit. Sure, ask questions and disagree if you must. In a collegial way. If a contractor says something that sets you back on your heels, it's better to make sure you really understand what he meant, have him elaborate, and then stick it in the mental file for looking into later on your own. One thing is true: prices do not drop when you can't keep a lid on it. Name-calling is never appropriate. Try: "I'm not happy with this."

There is a school of thought that it's a rough and tumble world out there, contractors are tough guys so if you can't stand the heat get out of the kitchen. When you run with the big dogs, you have to do as the big dogs do. That's do-do.

To put a shiny light on it, ethical behavior is manly too. Honesty is manly too. Taking the high road is manly too. Keeping promises is manly too. That kind of macho talk is for the purpose of manipulating people into doing the wrong thing. Never trade your power to engage the authorities on your behalf in exchange for a relatively tiny bit of money.

When you stick to above-board, document-on-paper modes of behavior, you will be lucky. There is nothing wimpy or powerless about being lucky.

Sometimes a symbiotic relationship can be worked out, some bartering as well as money, or to put it crudely "you scratch my back and I'll scratch yours." This isn't wrong, and isn't even evading taxes. It's the same as getting credit on your trade-in when you buy a car; you pay less, and no taxes are charged on the trade-in value.

This kind of deal requires trust, however, and is more appropriate after a few dealings with the same vendor, not the first time you hire them. Do not hand out trust like candy at a parade. Trust is earned over at least three jobs. If this deal with this contractor that you've never worked with before boils down to requiring trust, this is a red flag, warning lights, ding-ding-ding!

It's possible to do a barter on the first hire; simply write the details into the contract, do not keep it verbal. If you trade designing a website for having a new breaker box installed, it certainly can be spelled out in writing, even if no cash will change hands.

Contractors can be cranky. I like the cranky ones best, but that's just me. It's only the nice one, big smile, good talker, with charisma, who can be a swindler. I do not know of a single case of a cranky, irritable swindler. When the guy tactlessly snaps that my theory on the cause is WRONG, he becomes more trustworthy because he'd rather be good at his craft than please a customer. When he snorts "No way that can be done in three hours, there are at least four pipes that have to be rerouted (idiot)" he just rose to the top of the list.

What's even better, cranky guys know they were cranky so can be counted on to make it up to me with a lower price. Especially if I remain nice for the rest of his visit. Statistically, if there was a cranky one in the bunch, he's usually hired—due to knowing his stuff and low price, not mere rudeness. Cranky doesn't bother me. I grew up around old Germans. If it bothers you I'm sure some of the nice ones do good work too.

I had a cranky mover once. I wanted the whole move from one house to another house forty minutes away to be done in less than an 8-hour day, so I boxed up over 100 boxes myself and told him I would have almost everything from the second floor on the lawn come move day. He estimated it would take his crew until 1:30 to load the truck. I replied it should only take until noon. I said, "I told you, I will have the second floor on the lawn." He snapped "I've been doing this for twenty years and you're going to tell me how long my crew will take?" Well, he had a point, but I didn't like it. I hired Mr. Put-down anyway. One thing for sure, he wasn't an 'oh, whatever' kind of guy. I'm not an 'oh, whatever' kind of person either.

Come move day, instead of the usual three-man crew he brought a four-man crew, and was done packing up by noon. I paid the agreed-upon move price, no extra for the fourth guy. Guilt, and a bit of 'the customer is always right' policy at play. They worked hard and steady. Nothing broke, nothing was lost.

Being corrected for estimating too low happens to me all the time. I don't get bent out of shape, and I usually learn something new, a step I'd forgotten or why the process takes longer than it appears. Valuable information. I'll do better next time. Positioning myself from the very start as someone who is paying attention to how long things take works out well. Even if they correct my math.

Business Procurement vs. Home Improvement

Most people have heard a little about government bidding rules. These rules were devised to level the playing field, prevent side deals with the people contracting on behalf of their employer, and not reward industrial espionage. Yes, the latter happens and is real. In the high-stakes game of multi-million dollar contracts, all manner of theft and bribery can take place—and will, unless the repercussions are severe for people on both sides of the table.

But none of that pertains to you. In fact, borrowing buying practices from government and big business procurement departments to use for small fry household purchases exposes you to greater risks, not fewer.

The government rules are designed for situations where people are spending someone else's money, not their own. You don't need safeguards against embezzlement or kickbacks when you are buying for yourself. You need and want favoritism in vendor selection because you have to weed out swindlers, failing businesses, the inappropriate and the incompetent. In your case the contractor has the inherent advantage, so you have to play your side well. You have to trust your inner barking dog even when you can't put your finger on what he's barking at.

I've been involved with bidding and procurement under Government policy, Fortune 500 corporate policy as well as homeowner buying and selling. I know all three types pretty well.

To have a little fun, here are the differences in a nutshell:

The Big Dogs vs. Small Fry:

1) Big Dog: Let each bidder derive his own price. Small fry: Tell the competition the price to beat. The bidders want this. If he can beat it, he wins the work, and if he doesn't want to try, he stops right there. If he beats it, go back to your previous low bid to see if he will win it back. This is Contractor Heaven. It is exactly the way you use multiple bids to ratchet the price lower. Each contractor is given a chance to counteroffer; each contractor decides whether he wants to fight for the work with a lower price. A customer never cuts a finalist out of the running due to a $100 or $400 difference without giving him a chance.

2) Big Dog: Never mention details from the other quotes to the competition. Small fry: Take an idea you like in one bid and demand all the other bidders incorporate it into their quotes. This is simply making the quotes apples-to-apples, not apples-to-oranges. You must make your quotes apples-to-apples, or else how can you tell which is the better deal? If you had all day like procurement officials, you could adjust for the differences yourself; but that's silly when you can throw the task back to a guy who does this for a living. All bidders must bid on the same thing.

As you move up the learning curve, as you continuously improve your plans, you change your mind. The best time to change your mind is EARLY, and changing it before even hiring a contractor is the absolute best! Never be ashamed to find you forgot something or heard about a better approach; incorporate it! Email everybody with the new idea, and ask them to add it in. This is your life, your house. As you do your best to incorporate every good idea, communicating these thoughts before materials are ordered is ethical and cost-effective. Anything you leave out of competitive bidding, only to add later, won't be assigned a reasonable price and must be quarreled into a semblance of reasonableness.

3). Big Dog: Keep confidential not only the price, but even the names of the other bidders. Small fry: Tell all to friends and family. There is no secrecy around the bid a contractor sends you. Most businesses post their prices for all to see. Businesses live or die by their prices. The price is no more secret before you hire him than after. The government does what it does because it puts huge purchasing power in the hands of people making $80K a year. The point is to prevent *their own employees* from taking bribes to grease the wheels.

When shopping for yourself, there is no implied confidentiality for a quote; you may stand up in church and tell the congregation how much your new roof will cost. You may show all the quotes to your neighbor, friend, or handyman. You may tell co-workers all about them. You may post vendor name and price information on Angie's List—there are fields to fill in with that exact information. The myth of confidentiality harms buyers. It's difficult to get help identifying opportunism early if you don't share dollar figures with friends and neighbors.

When you shop for a TV you ask around, make some phone calls, read reviews, and google it. Maybe you don't even do that; with experience you have developed a trusted vendor or two and know a typical price before you walk in. There is nothing unethical about any of that. It's called smart shopping. You wouldn't buy a 21" LED TV for $1,400, would you? That was the price in 2002, but today that's a terrible price. Today, $200 is more like it, and $130 is great but what features did they leave off?

Home repairs and upgrades are no different. Where does this idea of confidentiality come from? From the unethical contractor himself. He simply will not make a dime if his victims talk to people. You will find him out. That's a good thing.

4) Big Dog: Start out with a level playing field. Small fry: Start out with favorites. Ask neighbors, friends and family for both recommendations and contractors to avoid. Listen to their stories. Send out RFQ after you have a mild preference for one or two contractors. It is important to have favorites because you need to speak to them second, third or fourth, never first. The first contractor you speak with will teach you the most but will also be the highest price; pick a shop you're lukewarm about to be that first guy. He's the discard.

Having a favorite is the way to ensure you get great prices from good contractors; don't ignore your inner wagging-tail dog either.

5) Big Dog: Send the RFQ, pick the lowest bid. Small fry: Said no one ever. The big dogs don't do this. Some people believe the federal government conducts purchasing by sealed bid and after 5 PM on the deadline date, opens them and gives the job to the lowest price. Not so. In April 2012 I attended a UID Forum organized by DoD procurement specialists and one of the trainers surprised me by saying the US Government hadn't conducted a sealed bid purchase for more than ten years, and possibly fifteen. Ironically, you may try to emulate how the big dogs do it and the big dogs think that's not a good way to do business.

The sealed bid process leads to all kinds of bid errors, omissions, problems with delivery, cost overruns and possibly going out of business before they could deliver. It was a kettle of worms. Buying complex assemblies and equipment by price alone is foolhardy. It could mean the least experienced firm won the bid only because they calculated one tenth of what was really needed for inspection or scrap disposal. Today the government considers a firm's experience, reliability and history of on-time deliveries, with price being forth if not lower. If a vendor isn't right for the job it doesn't matter what their price is.

The Government procurement specialist said they often conducted bidding for large things in several stages, narrowing it down to a handful of firms with the required skill set that scored well enough on quality and other measures. With a ballpark figure in mind, two or three finalists were asked to present a more in-depth proposal package with pricing. Hmmm, just like we should be doing with our household projects.

When you have several bids and one is significantly lower, it's better to just discard it rather than engage him in any discussion because swindlers live in the lowest bid. If you talk to him, he could make up convincing lies so you give him the job. Head in noose . . .

6) Big Dog: Must plug your ears to those dishing dirt about the competition. Small fry: Pay attention, tell me more. You must be self-interested to the best of your ability. Better yet, seek this information out. If you catch a whiff of rumors that one of the contractors is in financial difficulties, not paying his suppliers, uses a material that deteriorates in a few years, has high turnover making his workforce inexperienced, or anything that could make them less likely to do a good job, listen. Check it out independently. It's part of making an informed decision. It is a legal requirement for big businesses, called performing due diligence.

Be cognizant that the competitor as a lot of motivation to lie. Huge motivation. But on the other hand, contractors are just people, meaning not good actors and often prone to schadenfreude over the misfortunes of their competition. They are as likely to be the source of good intel on the competition as they are to be conducting an unwarranted smear campaign. Closing your eyes and ears to possible risks is not being ethical, it's just plain dumb.

What you should do exactly like the big dogs do:

1) **Adjust the payment schedule.** Big corporations and the Government set these, and their vendors take it or leave it. Sure, vendors will grumble or pitch for keeping their very wishful-thinking payment schedule, but should become amenable to a reasonable one in a minute. If they become annoying in their resistance to working before being paid, show them the door. Just like your boss would if you demanded to be paid up front.

2) **Dictate the materials and inspections.** You must say 'contractor to obtain all necessary licenses and permits' and specify the material you prefer. What you don't require in the contract may or may not be done; permits and Building Inspector visits are mandatory, so there should be no objection to clear language in the contract.

Case Study: Years ago a co-worker obtained seven quotes for a complex automated workstation to replace an old, worn-out version. He awarded the job to a company in another state, which was puzzling since these things need tweaking in and that would entail a lot of travel on our dime.

He replied he didn't want to, but our company had tried many drill spindles over the years and the only brand that worked in our application was Brand A. This was the only company who listed those spindles; all the others called out something else. He discarded their quotes, even though three were very good workstation builders with a thirty minute drive.

"For gosh sakes!" I scolded, "If you knew only Brand A would work, why didn't you just tell them all to use Brand A?" "Well, maybe they wanted to use another brand for some other reason, equipment integration reasons, you know." "Why on earth would they want to process-engineer and price components on multi-ton machine, then type up a six page quote that you're going to toss simply for using the wrong brand on a $300 component?"

His reluctance to making the quotes apples-to-apples prior to selecting a winner hurt the best contractor and hurt us. The notion that modifying the requirements with all the suppliers is somehow wrong is completely false. It can't be emphasized too much: the best time to change your mind is EARLY, and before signing the contract is the absolute best.

Most bidders will love being handed a component make and model, not have to leaf through catalogs to pick one. It's your right to tell your vendors the materials and finishes you want. It's immaterial if you decide after sending the initial quotes; just send an update.

What if you don't know the material you want? Then it's fine to leave it up to individual bidders to make recommendations. When you decide on a feature, say to have colored concrete on your new sidewalk instead of plain grey, don't throw out the uncolored concrete quotes due to being the wrong thing. Call them back with your color choice and ask them to rebid. It takes you 30 seconds and takes them a few minutes. Apples-to-apples.

3) **Accept no bribes, gifts or under-the-table deals.** A businessman doesn't do these for your benefit; he does them for his own benefit. How it plays into his hand isn't apparent to you yet, but don't assume 'there's no way this can go wrong." It can, and will.

They will give you notepads, refrigerator magnets, bag clips and whatnot with their logo; those are marketing materials. It crosses the line when it influences your emotions or makes you feel reciprocity is in order.

4) **Be up front about future work.** If this is the start, if you're planning to give a larger project to the guy who does a good job, say so. It's the perfect way to ensure that you get a good time-and-materials price and are happy with the outcome. The willingness of a company to be very pleasant and accommodating to their bigger customers is totally scalable. It's true for Exxon and it's true for your handyman.

5) **Solicit a basic bid to ascertain contractor interest, review and narrow it down to two or three, and have only those jog up a detailed proposal.** This is respectful of the vendor's time; only a few have to generate every detail, check on every materials price. You could start with bouncing your dealbreakers off of twenty vendors to weed out the ones who don't meet them. For instance, if you need a next-Monday start, and completion by end of week, phone to see which ones have a crew available or can do it. Making only the ones with a good chance of getting the work do a full quote is good. This is more important if you will tap these same guys multiple times in the next few years. You don't want to wear out your welcome. If it's a one-time deal, just make sure you get enough quotes to see what's going on.

Payment schedule

The payment schedule is your main tool for both weeding out the swindlers and keeping work progressing on schedule with regular businesses.

The buyer always sets the payment schedule, and the seller takes it or leaves it.

At first blush this appears wrong; we buy from stores all the time and they set the price and the payment schedule. What you're seeing, however, is the end result of hundreds of years of working it out, resulting in a system that suits the buyers because they won't buy otherwise.

Buyers pay the store when the complete product is in their mitts. The store can set any price they want, but the buyer decides when they'll pay and if they'll pay.

In the grocery store you pay for a can of corn when you have it in hand. But the farmer incurred his costs of growing it eight months ago, and the canner paid wages to their employees two months ago, and the trucking firm that brought it to the store paid for gas and driver three weeks ago. They did this without any pre-payment from you. No one said to you if you want a can of corn in December, give us 20¢ up front to cover expenses.

In money-for-work situations, both parties feel risk that the other party won't do their part. Businesses want payment before delivery, and customers want delivery before payment. Both sides have anecdotes about bad checks, deadbeats, fly-by-nights and bad faith. A business can mitigate some of their risk by vetting the credit card or check for the small down payment to make sure it's a good account. In addition they can avoid risky customers by highballing or no-quoting.

Statistically, the greater risk is on the customer side. A business comes to the table with experience while a customer is often a first timer. Therefore the greater protection should fall to the customer. That is why delivery before payment is fair.

In some states contractors may charge no more than 30% or $1000 up front, whichever is LESS. Your fellow may twist it around by saying the state requires 30% up front. Pshaw, baloney. The state would prefer 0% but 30% is a compromise against the abusive pre-payment demanded when ungoverned by law.

Down Payment: there is no legal or state requirement for even a dollar up front.

No matter how certain he sounds that it's mandatory, he's pulling your leg. Even if he shows you something on paper. To put a shiny light on it, laws related to contractors, just like laws related to banks, are about reigning in rampant abuses, not paying him unearned money, so any contractor citing one is putting a spin on it.

When speaking to the estimator, ask about the usual payment schedule and then mention any changes you'd like to make. The estimator might say "call the office, ask to talk to Linda" to ensure it gets memorialized on the typed contract you're eventually sent. The payment schedule may not be his department, but he will give you a name.

When their contract has payments getting ahead of job progress, pick up a pen, edit it, and both initial and date the change. A business that is inflexible on a huge down payment has something else going on, is either financially unstable or is up to something.

Don't pay for materials he reported as purchased but are not yet on site. These fellows are invoiced by suppliers and hardware stores, so unless they've ruined their credit rating they do not pay in cash. Paying them several days after materials arrive is plenty of time for them to pay the bill in full with no sweat or cash flow issue.

A good payment schedule means your job progresses and finishes on time. The primary reason no one shows up for days is payments ahead of work. Once he has your big wad of cash in his pocket, he really loses steam for doing actual work.

The chart is in percentage of the total, not payment amount at that point.

The cost of change notices are handled separately from the main contract: either all of them are payable at the end, or payable when the change notice portion is complete, depending upon the written agreement between the two of you.

A payment chart based on job duration is shown below.

	Percentage of total paid			
Project Length	1 to 7 days	8 to 15 days	2-8 weeks	> 8 weeks
Start	$0 or $100	$100 or 10%	10%	10%
Milestone 25%*				20%
Milestone 50%*		35%	35%	40%
Milestone 75%*			60%	60%
Completion**	90%	90%	90%	85%
Holdback, 30 days	100%	100%		90%
90-120 days after			100%	100%

*Milestones optional; use only when a distinct and appropriate point exists; use the actual percentage

** Completion means a sign-off by the customer

Milestones

Milestone payments meter some money to the contractor to help with his cash flow for expenditures, such as wages, on your job. They are keyed to completion of a part of the project, such as when an inspection is passed. There's no need for milestones in projects taking less than seven business days because the check can't clear before the project is done. *A guy who wants one for a short project is intending to not progress for weeks.*

Contractors work to milestones, so after getting a milestone payment, you can expect the work crew to be absent, off chasing a milestone somewhere else for a few days. This is rock solid if the payment is more than the work accomplished, less so if it lags behind.

This is why 'complete the teardown' is the worst possible moment for a milestone payment. If your job is going to stand idle at all, it will be immediately after the milestone is reached. Never locate a milestone where an uncomfortable or unsafe site condition exists. Never let payments get even with the work; always lag 10% to 20% behind. You aren't hurting him by doing this; he put a 20% or better profit margin on top of costs into his bid, so all you're doing is ensuring he sees his profit after the job is entirely done.

Ensure the milestones selected accurately reflect the completion level, and indicate the point with 1) a dollar figure 2) the milestone accomplishment; and 3) a percentage of completion.

Case Study: Say a contractor leads you to believe he'll attach the fireplace mantelpiece after the flooring is installed and all walls are taped, sanded and painted, so you make it the 70% milestone in the contract, at which point you'll write a check for $8,000. The job is progressing, and lo and behold the mantel is attached when only sawdust-covered plywood is on the floor, the drywall is half up and no painting is done and he wants his milestone payment. That is plain bad faith.

It's his prerogative to change the sequence of work for scheduling reasons, but that does not compel you to treat a job 40% done as 70% done. Since the mantel is up and the job isn't even half done, the contract requirement isn't yet met; let him know you respect his authority to determine sequence of work but a heads up would have been nice so we could pick another milestone for the 70% mark without this awkwardness. Or you could tell him nice try, ha-ha. On third thought, with your frowny face on, point out there were good-workmanship reasons for completing the drywall, painting and flooring before installing the mantle and you still want those, so you take this to be a quality reduction and wonder how he will compensate you for that.

Bank Loans

A word of warning about bank loans and very large projects: banks will release milestone payments by date even when the homeowner calls to tell them not to pay. Even if the contractor made no progress at all since the last payment, the bank will cut a check and your tardy contractor will have it cashed two breaths later. The bank can become an unstoppable enemy, throwing tens of thousands of dollars at non-performing GC. Money the homeowner can never get back.

The truth about banks is that most of their employees are high school graduates trained to just follow rules. Those who have the power to make decisions at banks are often too high up to bother with your issue. If there's a payment schedule, that money will be released on time; no verbal protest trumps a signed contract.

Banks can have different standards for loans, but one thing is certain: do not give them a calendar of payments. They will hand out your money on schedule even if the guy does nothing for ten weeks. One solution is to get your loan structured as a line of credit. You personally write checks on that line when appropriate and for the amount that makes sense. After the project is done you can refinance it another way for a lower interest rate.

A Good Payment Schedule

A good payment schedule has payment lagging significantly behind work—not just a few days, but say ten days—to keep your contractor working on your job.

If business is slow, his workers will be there every day even with a bad payment schedule. If he can get six other jobs and nest them between two crews, you will get a crew only 1/3 of the days, if you're lucky. In cases where one of the other jobs is twice the size of your job, you might see workers less than once a week.

If you write a fine for non-work days into your contract, there will be many days where a crew shows up for fifteen minutes and then leaves. The excuse will be 'gone to pick up materials' or 'gone to let the adhesive set' but it is really to avoid contract repercussions.

Case Study: One fellow acting as his own GC, to save money on house construction, wrote in his contract with the framer that the job had to complete within 45 days of start, with workers on site at least four hours per day until completion. They signed in March, and his plan was to move in before September. What happened was the framer never started the work, having overcommitted on other jobs. The basement was poured in March, ready for the framers, and by August, this contractor had still not started the job.

Before they finally started late in the year, he had to pump the water out of the basement twice, kids broke all the basement windows with rocks, their financial situation was shot and he had developed a new facial tick. Before his foray into 'saving money' by being his own GC was over, he was let go from his job for taking off too much and not having his head in the game at work. Sure, his firm was in the throes of some minor layoffs, but he was a manager, a group usually immune to layoffs.

The contractor truly does not care if you incur site damage because he doesn't show up. He is juggling eight jobs and gets his back up when you demand workers today. He replies with statements like "I'm doing my best." and "I'll get to your job as soon as I can." These are reasonable and businesslike responses to him. He feels he didn't promise anything, so the damage is all your fault, not his. This is how he sleeps at night.

He has signed too many contracts and now is simply trying to satisfy all of them without being sued. All he has to do is stave off customer complaints with some blah-blah to delay being sued another week or two, then perhaps he will be a bit more caught up.

When he takes on too much, material ordering suffers. He forgets to order things with a long lead time first, doesn't get the sequence right, or orders materials for five jobs at once even though for two of those jobs the materials will need to sit out in the rain for three weeks. It may be damaged beyond use—if you're lucky. If it's just slightly damaged he might try to use it anyway, causing huge issues five or ten years from now.

The Holdback

It is too high a hurtle to expect perfection when the job is complete. A good goal is quick response to your callbacks.

Don't expect the seller to volunteer a holdback in the contract he presents. Some do, but it's normal to remind them.

Do not pay the holdback just because the thirty days are up. The holdback is paid AFTER all issues are resolved or around thirty days, whichever is later. If you still have an issue open at the twenty-five-day mark, consider making a call or sending an email stating that if they don't take care of it you will hire someone else to do it and the cost will be deducted from the holdback, to the penny.

Either this email will prompt action to do the repair or they will say sure, hire someone else. The third option is they will be silent because they don't like either of their choices, but that's OK. If you're certain they got the message, and are just being nonresponsive, go ahead and get it done and send them what's left of the holdback plus a copy of the 'someone else' invoice. This is standard practice. Good communication is vital; if this escalates later on, your strongest case is that you reached them via two methods: mail, fax, email, text, or phone – and not an answering machine message either, talk to an actual person.

If the cost exceeds the holdback amount you may reasonably bill them for that amount. If that is unsuccessful, follow the steps in the *Getting Your Money Back* chapter.

Signoff at Completion

Signoff at completion is an acknowledgment that the work and cleanup is complete. It means the contractor is leaving the premises and assigning workers to other jobs. If issues arise with the work, the contractor will return to fix them; the signoff does not remove obligation for this.

The holdback gives the homeowner clout if something is still not to their liking or they suspect flaws may show up soon. Also, the signoff can be delayed for a week or two until there is more confidence all is OK.

If your contract requires a signoff prior to the completion payment, do not be intimidated into signing off when you feel cleanup isn't done or will be done tomorrow. Stick to your guns and don't sign off until it's really done. I'm not going to tell you the dozens of scary stories involving trusting homeowners and contractors who have been good up until now. Trust me, do not sign until it is really done.

When a one day or less installation is involved, such as a dishwasher or refrigerator, the installers may present a piece of paper to sign that has a bunch of legalese saying the homeowner is vouching that everything is OK. The savvy homeowner feels the urge to balk at signing since it's not clear yet whether it truly is OK.

In general it's fine to go ahead and sign if the product and installation has a warranty. For practical purposes it's a billing sign-off. This doesn't negate the warranty. Most of the time they want happy customers and simply require an acknowledgement of the day and time the work took place. The legalese is regrettable, but for the most part it doesn't outweigh the manufacturer's guarantees of satisfaction. If you want to, and I've done this, write a little note saying 'not tried out yet' or 'condition not yet verified' and then sign.

Contracts

A good contract deters legal action. The contract has failed you if the other party has fallen short of expectations yet they feel they could win in court. Since the seller mainly cares about safety, convenience of his workers, legal protections for himself and getting paid, not in that order, his contributions to the contract will dwell upon those. It is up to you to add what you care about.

Contracts which cover only the high points are risky. If everything goes smoothly it works fine. When it doesn't, it will be because your expectation wasn't mentioned. Oddly enough, it's a stronger legal case with no written contract at all rather than a skimpy one, because once there's a written and signed contract it becomes the four walls of the agreement and nothing unmentioned can be considered part of the deal.

The judge shows on TV such as Judge Judy show how oral agreements are adjudicated in court. The shows function as small claims courts, and like the ones in your county, base decisions on which party is more credible and which side of the story harbors a smaller amount of the blame. But the judge's opinion serves only in the absence of a signed contract. If there is a signed contract, no matter how skimpy and inadequate, it is the full agreement and nothing outside of it may be admitted.

One option, when faced with signing a skimpy contract that kicks off a project where much is unknown or unspecified, is to add a handwritten sentence or two that makes your review and input at certain points part of the deal.

Some sample lines might be: the buyer will provide written instructions for (XYZ); customer must sign off on materials before installation can begin; the sketch of the design shall be signed and dated by the customer before framing begins; work schedule in writing and initialed by both parties prior to start and notification of schedule change sent by email twelve hours prior to that change. Input instructions like this punch big holes into the four walls of the agreement. That returns you to a spitting contest of who did what and when, but at least you have a chance of prevailing.

There is no rule about which side gets to write the contract. The other side can, or you can.

Being convinced that your wants and preferences have just as much right to be in the contract as theirs do is 90% of the battle. Their contract is a starting point; any clause can be crossed out or amended.

Even if they use the three-carbon kind with the pink and yellow pages behind, it is editable with cross-outs and margin writing. The tiny writing on the back can be edited too.

In practice, because the seller does this all the time, he has a standard contract. Because he wrote it, all the breaks go to him and all the grey areas lean in his favor. If the job goes uneventfully all is fine. Should anything go the least sideways on this job, however, you won't like the outcome.

You could bring your own contract and ask them to sign it. It's not out of bounds or unreasonable. To make it easy there are good contracts for sale on-line such as the AIA (American Institute of Architects) Contracts On Demand.

If you have some edits you'd like to make upon seeing their contract and the seller presents a reason why the contract can't be edited to include elements which protect you, that should give you pause. Reasons like "company policy" and "our legal staff won't allow any changes to the basic contract" are meaningless. They're big boy ways of saying "I don't wanna."

The company representative must play for his team, and by stating it's company policy that the contract is not edited by the customer he has done his part. Now the discussion is 50% done.

There's no doubt at all; if you're handed a contract by the other party that's missing details and features, or has clauses that he says don't apply to you, yet he will brook no edits, walk away. He is up to no good.

It's always OK to ask for time to read it over at home before signing. It's your legal right.

Contract format usually has two parts: Boilerplate and body. The body contains the details about what you are buying or getting done. It's located on the front side of the paper, or top of the only page.

Boilerplate is on the back or on the bottom, often in smaller text than the primary or body part of the contract. Boilerplate is standard protective language that covers seldom-occurring situations, provides uniform due dates, timeframes and standard practice for that business.

Most of the time since they have made a good first draft of it, just go through their contract and edit it. It's less work all around to edit it and email the revisions to them for review. With small businesses, when saying I have some edits to the contract, an office staffer will offer to type them into the contract for me, rather than have them as margin edits. If they offer, take them up on it. Then compare it carefully to your notes for typos and missing parts.

An adequate contract contains:

Names, addresses, phone numbers, email addresses of both parties

License number, insurance carrier(s) of the contractor, policy numbers

Total price. Several itemized prices as appropriate

Hourly rate for additional work beyond the scope of the contract[14]

Payment schedule, with conditions for payments

Dates and duration

Details about products, work, brand names, numbers; as much as possible

Details about cleanup, rules, owner requirements, leftover materials

Holdback duration and amount, conditions for paying

Permits purchased by the contractor. Ideally, copies are attached

Contractor's work guarantee (usually one year, stating what it covers)

Directions on handling amendments to contract

List of subcontractors and suppliers. It's fine to limit it to those providing over $200 in goods or labor

A requirement for lien releases from all subcontractors and suppliers listed

Signatures of both parties

Addendums (including a copy of the email or request with specs and details)

 There are at minimum two Addendums to add to any contract; one contains your list of expectations, specs, mention of any products or brand names you wish used, verbal statements the estimator made, and two, the RFQ you sent out.

 If the RFQ is not an Addendum in the main contract, whether your contract or theirs, it is non-binding and nothing specified or stated in it will be part of the agreement in a court of law. If some things changed, still add it but make a note in the other addendum to supersede it. Both parties initial all Addendums.

---

[14] Contracts for one or two-day jobs, such as painting, tile-laying, sump pump replacement, fence repair and so on may not need hourly rate, amendment-handling information, or permits.

Add things like a magazine article with a photo you want yours to look like, or copies of emails you sent asking for certain features, as Addendum C, Addendum D, etc. Date each one. Both parties initialing and dating every page can be a good idea if there are changes from the RFQ. Adding in documents the other party provided, such as brochures, is a great idea if something in there is included. Circle the relevant bits.

Non-disparagement clause. This clause forbids you to write a bad review online, even on sites like Angie's List. Contractors put it in because unethical people and nutcases are out there. If you read enough reviews you'll spot them. Compelling no complaints, however, is of dubious legality. He should be dealing with problems by fixing them.

This book isn't aimed at consumers who indulge in illegal blackmail, so my assumption is if you feel moved to write a bad review, it's because the service and work were exactly as stated. Hating to get bad reviews so contractually forbidding them steps over the line. Any clause that denies you recourse—can't sue, can't arbitrate, can't complain, sets a ceiling on recompense—is out of bounds.

Substitution clause. This clause allows him to use whatever he wants—nearly always something far cheaper—when the contract specifies items by manufacturer, model, type or number.

The rationale is that if a product is unavailable, he can substitute something else to keep the workers busy and the project progressing. When you have called out an exact model number, it should not be easy for him to invoke this. Either cross it out entirely, or change it to require him to try three sources prior to calling it unobtainable. To invoke the clause he will need to provide those names, websites and phone numbers to you in a change notice for sign-off, plus list the model, type, etc. that he intends to use. It goes without saying that some cost change should occur; if he isn't charging you more because it costs more, then he is pocketing the difference because it costs less.

Arbitration clause. Check out who the assigned arbitrator is and the scope of their powers. Check out any limitations, such as no settlements beyond the actual out-of-pocket costs. Arbitration can be great–such as when it is performed by retired lawyers in the evening at a location four miles from your house and is limited to double the full project cost. Or it can be designed to be a huge pain in the neck, such as when required to be performed by a group which is in the pocket of his professional organization in a location 90 miles away, held only once a month and limited to 'damages' which means a cost that shows up on a bill.

Example of being limited to damages: if the contract is for a $700 high R-value window but he installs a $400 leaky window, the owners get only a $300 check if they win. This arbitrator can't compel him to install the high

R-value window. Nor can she throw in punitive damages to compensate for future higher utility bills.

If you read about arbitration, reviews will range from 'wonderful' to 'a nightmare.' Venue, who conducts it, how impartial, how knowledgeable, and the size and nature of the settlements they are permitted to assign makes all the difference.

Arbitration is often quicker than the courts, and the contractor has pre-agreed to not contest the ruling. He has picked the arbitrator, which means he feels confident there is no bias in favor of the customer. Is there bias in favor of the contractor? That's for you to find out before you agree. If you don't have the time, just cross it out, and both initial and date the cross out.

Keeper of the Verbals

Verbals are those comments or promises made during discussions about the project, when the estimator says 'oh we always do that' or 'we use the top selling brand name' or 'we're slow right now so can deliver in three weeks if you sign by Thursday' or 'we won't need to drive over the lawn.' If there is a written contract, every verbal must be memorialized in writing or it is not part of the agreement.

A oral statement is not binding when a written and signed contract exists.

You are the Keeper of the Verbals. Misfortune arises from assuming the estimator has his sales pitch written down somewhere and will ensure it gets done as promised if they win. He doesn't keep track of everything he says and may not convey anything about the visit to the office staff—he doesn't even know yet if he has the job. Even with good intentions, people forget. Without a written reminder in the contract that the office staff can transfer to the Supervisor's job packet who then conveys it to the worker during the morning work assignments, it might not happen.

In addition to keeping all of his verbals, write down your verbals too, all the things you asked for and he agreed to do. Next week when he's fleshing out the contract he's not only going to forget what he himself said, he'll definitely forget what you said.

Companies change suppliers all the time to get better pricing, so if your decision was swayed by use of a certain brand name, state it in the contract.

When discussing your project with the estimator, have a small notepad and pen in your hand and note down every the verbal feature, process or type, including things he says is standard practice or how he always does it. Nothing is too minor to include in a contract if it's something you want to see happen.

When you ask for a verbal promise to be put in writing, the answer should always be yes or sure.

Take your notes, straighten them up, and provide the list to their office person for typing into the contract. Most firms will be good about this. If not, they can be added with pen to the basic contract at signing. Write in the wide side margins or wherever there is space. Both parties initial and date each note or groups of notes.

Another option is to make an addendum out of them. Put all of them on a separate sheet labeled as Addendum V. In the main contract write "See Addendum V for contract requirements and modifications." Have both parties initial and date each Addendum as well as the main contract near the note. This means that if the main contract and your Addendum disagree, your notes rule.

Write in full sentences. It is not more professional to write in bullet points or partial sentences. Too few words cause problems. Quibbles about interpretation can eat up days later on.

Writing the Contract

The vast majority of contracts are small and written by one of the parties. Bank loans and auto purchases may be the only time you see lengthy and complex legal talk. There is a significant movement in the law industry to write all legal documents in plain English. All the 'party of the first part' and 'hitherto' legalese is not necessary. Far from being required, the legal community is phasing it out.

Do not feel it is necessary to write a contract in unfamiliar language; write your meaning as clearly as you can state it with your normal vocabulary. Write in full sentences and try to include all your thoughts. Brevity is not your friend.

One useful thing is a little definitions section where a word, such as 'seller' is assigned to a name and address, ditto for 'buyer' and then in the document there are no pronouns, meaning he, she, her, him, they, them references, but the defined words inserted. It can be contractor and homeowner, or Sam and Melanie, as long as it's defined in the beginning. This makes it sound stilted and not plain English, but fairly easy to write. Brief descriptions of any codes or specs referenced are useful too.

If you get a loan to cover the cost of your home improvement using your principal residence as collateral, you have a three-business-day Right of Rescission. This means you can change your mind, back out of it. The provider of this loan should provide this form. A copy of what it might look like is in Addendum A.

Write the contract to stay out of court. Lawyers love writing contracts with all kinds of restrictions, rules and shalls . . . but not a single if-then. Why? Because when any clause is violated they prefer going to court to collect damages. The cost of that may be higher than the disputed amount. Instead, write your contract with the consequence right in the contract. If you do this, then here is the penalty. A well-written contract has the repercussions defined. Such as: If you're late the fine is $50 per day, deducted from the completion payment. Or another: Misaligned tiles must be removed and reset or a 3rd party will be hired to fix it and the cost deducted from the invoice.

If the repercussions roll directly from the acts then there are no violations, just movement within the contract. If you have a rule and can't dream up a fair repercussion, think about how that clause could be enforced under any circumstances. This will prevent you from clogging up your contract with petty, unenforceable rules. Carrot and stick. Not everything has to be stick.

If something is a true requirement, don't bolster it with a fine. Present it as a dealbreaker. Attaching a fine to a deadline or requirement doesn't mean the contractor will always dive in the right direction. If it's his busy season and he's nesting jobs, it may be an economic choice to incur your $50 a day fine to make $300 a day on another job.

A bit of legalese regarding in-contract consequences: a fine must fall within a compensatory range and not be punitive or usury or else it is contestable in court. A fine of $50 a day for each day of no progress on a kitchen remodel is compensatory if the kitchen plumbing or electricity is nonfunctional so a family of five must dine out, but probably not for a single person.

Fines can backfire; imagine a situation where a supplier delivers three weeks late, so the fine adds up to more than the balance owed. No contractor will complete the work for free. Writing the contract to exclude delays beyond his control, however, will mean every delay will be due to a reason beyond his control.

A better option may be writing into the contract that the contractor must give notice a certain amount ahead of time, or within a certain amount of hours upon becoming aware of the issue, plus provide a duration, to prevent the fine clock from starting to tick. It becomes a fine/communication/fixed timeframe combo.

Most of the time putting the disincentive right in the contract prevents issues, but not always. When it doesn't, the built-in repercussion can mollify your resentment, since you are made whole for your inconvenience.

When writing contract sentences, encompass the thought in a single sentence. This may mean that part of the prior sentence is repeated again in the next one. Each sentence should have a clear topic and action. In normal

English our sentences flow, and we assume we're still talking about the same thing as the last sentence. No such assumption exists in contracts.

Amend your contract: don't verbally discount part of it. If a direction or requirement was worth having in the contract when you signed it but now you don't care, don't give the worker a pass to ignore it. Amend it.

If you start down the road of winking at your own contract you'll regret it. Doing this can void other clauses in the contract that you had no intention of being affected. At best, it could lead to the contractor believing the nice guy act has no end, or that the permission to overstep was a blanket one for the remainder of the contract, when actually the customer wanted to bestow it magnanimously each time. Giving permission for a cigarette butt to be dropped on the lawn, a loud radio, or not putting a tarp over a piece of furniture while the room is being painted WILL be taken to be permissible from now on.

You must require your contractor to submit amendments also. If he changes a supplier or sub to a different one than listed in the contract, for instance decides to buy his lumber from a different place, he must submit an amendment prior to the lumber delivery. If he doesn't, simply does the ol' switcheroo hoping you won't notice, this is a red flag and may signal it's time to fire him right now. This is not innocent; a bona fide contractor gets quantity discounts from suppliers. He gets the best behavior from subcontractors who know he's a source of steady income. If he's ruined that with either of them, usually by being in arrears, he will try to hide it from you. If he's making a sound business decision to switch lumber yards because quality dropped at this one, or to use a tile layer who doesn't arrive for work inebriated, it's no problem to comply with contract and tell you.

It would be nice if contractors wrote up little papers and gave them to you in a timely fashion. In real life, what happens is a contractor tells you his changes, after which you write them down and then get him to sign. This is how it will go more often than not. Rather than resent this, rejoice. It means your understanding of the change is the only one that is approved.

The Law Loves Handwriting

The Law loves handwriting. The law loves it so much that in most states a final Will that is entirely handwritten, and 'entirely' is the operative word, doesn't require witnesses to be valid. It is perfectly legal, supersedes and renders null and void any earlier will generated by a most expensive lawyer with sober witnesses. Also in most states a handwritten will amendment, called a codicil, may amend or modify a section of a typed will.

The Law loves handwriting. When a contract has a printed section saying one thing and a simple handwritten note squished in the margin saying another, it becomes a matter for a judge to decide and nothing less will do, unless . . . both parties have initialed and dated the little scribbly note. Then it overrules the typed part in any court in the United States. Even if the typed part isn't crossed out.

Handwritten means in your best possible longhand handwriting. Even if you can muster up a crisp drafting hand, don't use it here. No printing. Connect your letters. If your normal handwriting has few connections, fine, just use your normal handwriting. If your normal handwriting is illegible, make a champion attempt at readable. It's better if one of the signing parties write it. A note in a different handwriting than one of the contract parties could be risky in court.

It goes without saying that all writing must be in ink. Good practice is to bring a blue, green or purple ink pen with you to the signing to make the additions distinct.

Never write in pencil on a contract. Pencil is a spoiler; erased pencil can have the effect of putting the legality of the whole contract in question. If you make a mistake in pen, cross it out with one single line through it. Obliterating with several scribbles is like erased pencil, risking questionable legality should it go to court. Just one line, a cross-out.

When both parties have initialed and dated the cross-out or handwritten addition, that note becomes THE MOST BINDING thing in the contract. Not equal to the typed text, but higher ranking than the printed sections.

Contracts historically have wide margins for exactly this purpose. It is not irregular to modify contracts with handwriting; it is a long established custom.

Most long contracts call for an initial and date box on the same page as the full signatures on the contract, plus call for initialing and dating each page because there's a legal assumption that some writing will occur on the contract before signing. Since most people write their initials very differently than they do their full signature, putting a set of initials on the last (signed) page ties the initials to the full signatures.

If doing this by fax, mail or email, generally I'll make the note, date and initial, then make Date _____ Init ____ fields to remind the other party where to initial and date.

In many states a contract is legally binding with one signature on it. It's the legal version of 'one cuts and one picks' that parents might do when splitting cake between two children. If one party entirely writes the contract and the other party signs it, it's considered the same as if both parties signed. But if both parties have contributed some words or edits of any sort, you need a copy with both signatures.

What if word choice in the handwritten note isn't great—will it be worse than no note? It's plausible, but I do not know of a single case.

The reason is that a good contract prevents contesting in court. If a certain oral promise is missing from the contract, the contractor can later say it wasn't part of the deal, but if there's a handwritten note on the contract that both parties initialed, however badly worded or misspelled, a judge would get the idea and the benefit of the doubt would fly to the customer.

At this point the contractor would have to prove to the judge that the note was about something completely different, not what the customer says it's about.

Ask for a copy after signatures, instead of getting a twin unsigned document when you walk in the door like the mortgage companies are fond of doing. Having a copy of your edits gives you more leverage later, motivating them to comply with contract.

The biggest misconception about contracts is that the legalese is so complex and everything's binding and precise down to the last comma and apostrophe. In reality, typical contracts have typos, are likely to have parts that aren't edited to fit this particular situation, and often contain sections that don't apply at all to this deal. It all boils down to what's worth contesting in a court of law.

You can edit any contract. Contracts are a meeting of minds, and your mind counts. Never sign when there's a part with which you don't agree.

Will a seller do things to get back at the customer who caught them trying to slide a major change through? Unlikely.

When a contractor sees you block his 'forgetting' to add in a feature or task in a nifty fashion, that makes you formidable. Now he knows you are on the ball and have a strong contract. He's not going to poke the bear anymore. This might mean the remainder of your relationship is polite and contract-compliant. Odds are good you get the best work and no funny stuff. Everybody smiles, everybody talks in calm voices, all is polite.

Cross out what doesn't apply. Add edits to the margins.
Both parties initial and date each one.
Have one set of initial and dates on the signature page.

Will, Must, Shall

In legal usage these three words have different meanings. 'Will' means it is compelling, 'must' means it is very compelling to the point of repercussion if it's missing, and 'shall' has no wiggle room at all. A politician said on TV, during his sound bite about the high cost of a bill before Congress, "We can't accept this bill. It has 186 shalls." It was a revealing statement if you know the real meaning.

To illustrate, an order for fifty hammers that 'will last for five years of daily use' can be bid at about $30 each. An order for fifty hammers that 'shall last for five years of daily use' requires life cycle testing of every tenth one produced, with the test equipment, test labor, test data analysis & scrap to be amortized into the cost of the hammer, and every potential weakness fixed with materials more robust than wood and carbon steel, resulting in a hammer costing $600 each. That's what the Congressman was saying about too many 'shalls.'

Avoid using the word shall whenever possible in legal documents. It bites both ways. It can compel a level of extra cost you don't want as well as make the smart vendors no-quote. It goes without saying, never use it accidentally because it sounds legalese.

* * *

Although there is much more to contract law, the usual small business or home-drafted contract can be good enough if the full thought is within a single sentence and pronouns, it-he-she-they, are skipped, dates and timeframes are clearly expressed, and ALL expectations and verbal promises are noted or referenced in an addendum. The goal of a well-written contract is to discourage the other party from contesting it, not perfection. When the other party feels they can't win in court, they will comply with the terms or skip town. In most states the courts charge the court costs to the loser, so when the other-guy's odds shrink these fuel their willingness to settle out of court.

Most of the time when a business relationship gets to the firing or lawsuit stage it's because one party tried to be understanding or let little stuff go, and these grievances festered or piled up until they hit the tipping point. It seldom is just one big thing.

Licensed, Bonded and Insured

Licensing is done by the state for contractors in many trades. During the estimator visit ask for the license number, which can be used to look up performance history with the state and the BBB. If your state licenses this trade, limit the contractors you'll hire to those having a license. Go online to the state's website to find the procedure for checking the contractor's license number. Then do that, to see if complaints or problems are reported. It goes without saying, be sure to report your problems if any arise.

In several states, the advantage to working with only licensed contractors is your ability to tap the pool of money the licensing board maintains for the purpose of making customers whole should the contractor botch the job or go out of business. In New York City, the maximum per-incident amount is $20,000. Other communities have varying amounts, or none, but the licensing boards will take complaints and apply some muscle on your behalf.

Many areas have minimum experience requirements or even tests which must be passed prior to approving the license. At minimum, they require proof of insurance. All good for you, the consumer. To find out about your state or local licensing robustness, search online for [**state**] **contractor licensing.** Look for the sites with **.gov** at the end.

Being licensed doesn't guarantee satisfaction with the work, for several reasons. Unknowledgeable customers can expect things that aren't possible due to physics or the limitations of the chosen material. Champagne on a beer budget, as they say, means the foam head is unavoidable because you get beer. Estimators should take their role of managing expectations seriously. If they catch whiff that the customer's imagination is outstripping physics, they should speak up, but often don't. Two, some personalities just don't mesh. When an antagonistic situation develops, small things can be blown out of proportion. Meaning a situation that would have been tolerable with a 'nicer' worker is now intolerable.

Bonding covers anything you might bring suit for, such as damage or theft. It also guarantees the contractor will complete the job in a satisfactory manner or else the bonding company will provide payment to you to get the job done properly and completely. Electricians and plumbers must be bonded, meaning their work is guaranteed to be good. Other trades with hazy grey areas between competence and personal preference are less likely to be bonded for job satisfaction.

For both bonding and licensing, don't accept the number or document as sufficient proof; check: 1) that it isn't expired; 2) it covers the trade or work he will be doing for you, and 3) when it was first acquired. Less than a year ago is risky.

A license bond guarantees the licensed contractor will comply with the codes and regulations set forth by the town, village, state, or other local ruling board. Government entities usually require these as well as contract performance bonds because their projects absolutely must happen and must happen for the agreed-upon amount, so insurance covers any cost-overruns and vendor failures to deliver. The cost of the insurance, which can be high, is worked into the price.

It is too expensive for the average homeowner to request a contract performance bond. When one is requested, the underwriter looks into the vendor's past history and the complexities of this project and comes up with a dollar figure, adding a few weeks to the process. It goes without saying that the better the contractor usually is, the lower the cost of the bond. An awful contractor will be uninsurable.

While it sounds like a nifty way to have an insurance company vet your contractor, don't yank the chain of an insurance company by having them do this legwork and then say 'never mind.' It's a bit more complicated than stated here.

Prior to signing the contract, request proof that the business is insured. Do not hire anyone who dodges this question. Ask for proof of workmen's comp and health insurance for his workers in the event they are injured, plus insurance covering liability and damage to your property. If the contractor doesn't have these, your homeowner's insurance would pay their claim, but doing so could cause a rise in annual fee for several years. In the case of contractors in your home, the business needs to have insurance for their own workers so you never have to deal with injury or property damage insurance claims.

Lien Release

Most horror stories about contractors involve them abandoning the job, either suddenly, or slowly over months. The reason is nearly always about money. The best tool to prevent you from having a horror story is the lien release.

The initial quote may not indicate whether he is subcontracting any of the work, but you can be certain he has suppliers for the materials. For shorter projects, state in the contract that you require an **Unconditional Waiver And Lien Release Upon Final Payment** from his subcontractors and suppliers prior to making the completion payment[15]. It's a good idea even if there's no subcontracting but materials are a significant expense. Limiting it to suppliers and subs over $200 is reasonable.

For longer projects with several milestone payments, write it into the contract that lien releases will be needed from suppliers and subs for work and materials up to that point. The lumber delivery lien release should be in your hand before paying the next milestone payment. If the flooring is complete, the flooring guy's lien release should be handed over prior to paying the next milestone.

A detailed article detailing how the law views liens is at: http://www.angieslist.com/articles/what-lien-release-or-lien-waiver.htm

---

[15] Completion payment or final payment is not the holdback, it's the payment before the holdback. Never hand over your holdback before you've lived with the completed work at least 20 days.

Contracts

A lien release is a legal document, available for free on-line,[16] that the contractor/subcontractors/suppliers sign saying materials are paid in full and no more is owed. The language and scope vary by state. Hang onto this piece of paper for several years. It sometimes takes a few years for creditors to start suing the contractor's customers.

Why is this necessary? When there's a cash flow problem a company stops paying its suppliers. They do this long before they start cutting payroll, so the employees, even the estimator, may have no idea. An unconditional lien release means if the contractor owes his suppliers any money, they have no legal leg to stand on to ding you for payment.

Why would a homeowner who paid in full care whether the guy he hired pays his bills on time? When it's work on your house or property, believe it or not, this third party whose name you never heard of can come after you to get their money.

Here's the legal thinking: Say you hired Basements Be Us to panel your basement family room, paid in full with a check, and are pleased with the completed work. Basements Be Us bought the paneling for nine jobs that same month on credit from Panel King and then paid only thirty percent of the amount owed before going out of business. Panel King can come after you to pay for the paneling because you and Basements Be Us were contractually-tied customers and you have the paneling. A year later PK tracks down eight of the nine BBU customers and goes after them for the 70% still owed. Without an unconditional lien release your recourse is to sue BBU, who took your money but didn't pay their bills. After you pay PK you can file to be one of BBU's creditors in bankruptcy court. It might take a long time and you'll probably recover merely a small percentage.

When Panel King sends you the letter demanding payment, fish out the unconditional lien release they signed and celebrate. You dodged the bullet. Your entire paneling order fell in the 30% that was paid in full.

Why don't all good contractors automatically give the lien releases at the completion of a job? Good question.

Lawyers may quibble that the above is too simplified, omits key words used in formal descriptions, and has not taken every variation on a theme into account. My view is that requiring this legal document and getting it prior to paying in full spurs greater contractor conscientiousness in paying his subs/supplier bills that month. No supplier is going to sign such a document unless they truly have been paid. It's akin to a receipt saying no more is owed, they're paid in full. That's the value of it, rather than being bulletproof legal coverage later. Getting a small pile of unconditional lien

---

[16] Addendum A contains copies of several lien releases plus the website to download your state's lien release form.

release & waivers from your contractor's suppliers and subcontractors means no hassle later on.

Another small plus to having a lien release signed by PK is warranty. If BBU goes out of business owing PK money, PK may decide to pursue it in bankruptcy court rather than exercise its legal right to hound customers. PK will, however, take a dim view of honoring legitimate warranty claims on what feels to them like stolen materials, if one defines stolen as 'taken without paying.'

Contract Omissions

When what you expected and what you're getting is a mismatch, 90% of the time it's due to contract omissions.

Normal businesses are surprisingly good at complying with contract. Many go farther than that and made sincere efforts to do the full job most people expect, even without detailed language in the contract. For instance, a contractor's usual contract may say nothing about doing cleanup or disposing of scrap material, but they do it. If they didn't, however, you couldn't enlist the help of the law. Your recourse would be in the realm of some of the steps listed in *Getting Your Money Back* chapter.

Hiring a lawyer will not provide much protection. The biggest problem with home repair contracts is omissions. If your lawyer wasn't present during every discussion, he won't be able to spot the features and processes that were agreed to verbally but are now missing from the contract.

When the contractor springs a charge on you for a sander rental he says was not included in the contract, your lawyer will not have caught the lack of mention of a sander before signing. If the contractor is silent upon rental equipment, a lawyer will not, on your behalf, dream up a line like "All equipment rental deemed necessary is included in this price with transport, safe handling, and insurance the responsibility of the contractor."

Frankly, if the contract is short and mainly just a list of services and products, there's no meat for the lawyer there. The only thing to look for is omissions.

Only YOU, knowing what you want, what was discussed, and how you define a good job can vet that contract to make sure it encompasses all of those things.

A lawyer is little help in spotting less-than-optimal processes, incorrect milestones or inappropriate materials which are memorialized on the page, unless you find the rare one who worked in that trade during summers in college. He is also no help in determining if this is a good price or if you are being overcharged, unless he researched a similar project for his own home in the past several years. In either case the help he would render is knowledge of the trade, not law. That's a high price for the help, considering that if you polled friends, relatives and neighbors you could find the same depth of experience in people who would love to bend your ear about it for free.

Lawyers miss harmful clauses in contracts all the time. If you do engage one, help him by pointing out whatever looks funny for any reason.

Omissions are the main source of homeowner disappointment. Get all the verbal statements, brands, and ID numbers into the contract because it prevents mistakes and makes your job go smoothly.

Watching the Work

To achieve satisfaction with your project, you need to master a critical skill: the frowny face.

Giving a frowny face means to make your facial expression obviously displeased, congruent with your message. Many people are unaware that they grin or grimace when they feel anxiety, and giving bad news to people causes anxiety. Often the frowny face can serve the purpose on its own, without words. I recommend putting on the frowny face three seconds before expressing displeasure. That way you can give forming the face your full attention prior to saying any words. It lets the other guy know what's coming.

All corrections and displeasures must have a frowny face. No weeding through a mixed message, which believe me, blue collar guys hate. People with English as a second language have even more trouble with it. People who work with their hands tend to be visually oriented, not aural; when their eyes and their ears disagree, eyes win.

Not using the frowny face but smiling throughout a demand for improvement can cause it to be interpreted as not serious. Then it's not rare for things to tumble down into firing or suing. Is that nicer? If you can't do a good frowny face, practice pulling your eyebrows down and scrunching your lips together in a bunch in the middle, as if holding a straw. It suffices.

Good frowny face

Very good frowny face

Not a frowny face; will be mistaken for joking around

Good frowny face   Excellent frowny face

Another fine point that will help your project: eliminate name-calling. This is a toughie, because people who name-call others are unaware they are even doing it. They believe they are 'emphasizing the point' or 'getting their attention,' when in reality they are bogging down communication. Name-calling provides no information on what is amiss and what is the preferred behavior.

Replace the name-calling with a direction, instruction, or description of what you prefer they do. For example, can you tell what I'm doing wrong and what the speaker wishes I would do when they say the following? "Are you an IDIOT?!?" "What the F—are you doing?" "How stupid can you be?" "You have got to be the dumbest (fill in nationality) I've ever met."

The above are a time-killing waste of breath. They slow down the corrective action. Statements like the following will bring you immediately closer to seeing the bad behavior fixed: "Hey, no dragging the ladder over my floor!" "Two guys lift that, not one!" "Read the can! It says no shaking before use!" "Get your butt over here and give me a hand!"

If you have trouble articulating the corrective action sentence, simply put on your frowny face and say "Hey!" or "Stop!" That should provide a pause, giving you time to formulate a useful sentence.

Living Through Renovations

If the renovation will be major, one way to save your sanity and dignity is to borrow or rent a camper, park it in the driveway and live there through the worst of it. The big plus to doing this is you still get your mail, the kids go to school exactly the same, there's a possibility of daytime privacy yet you are still on the premises to oversee the work. The minus is the contractor will feel since you are comfortable, he can delay ordering materials, stick air time into the job, and delay your job to send his crews to work on every single little job that comes his way.

Your tools to prevent this from happening are the Payment Schedule and your phone. Make calls if what you expected to happen today isn't happening, and don't wait until 4 PM to do so.

Before work starts:

Remove wall hangings. Work that shakes the walls, causes vibration or produces loud noises can cause items to walk off shelves and wall hangings to jump off their hooks.

Remove valuables from the workspace. Anything you don't want damaged by being handled, elbowed during work, or run over with a filthy dolly, remove to an area or room far from the workarea.

Remove breakables from shelves and mantelpieces anywhere in the house. They will walk off the shelves once the hammering and sawing starts. Sound is just vibration, and loud sound is big vibration. Put all your small things, breakable things into a room that is uninvolved with the project at hand. Close the door and post a sign saying 'Stay out.' Find out what language(s) the workers are fluent in, and write 'stay out' in that language right below it. If all rooms will be affected, wrest an agreement to work half the house then the other half, so you have someplace to call your own.

Remove all little throw rugs in the path of the workers. These things are stumble hazards for strangers, even though it never happens to you. Workmen are often carrying large, heavy things while not lifting their feet properly. Don't wait until they drop what they're carrying to break their fall before you remove that rubber-backed rug you were sure would not be a problem.

This contractor is not napping. If your contractor is laying down or slumped over with a tool near his hand, call 911 for an ambulance.

Pets

Pets should be off the premises during the demolition and work.

Have them visit the neighbors, parents, or anyplace else that is far away from the noise, dust, sharp edges and open holes that home improvement creates. Simply putting them in a room and closing the door will not work. Trust me on this. Dogs will bark the entire time, making it difficult for your crew to think—therefore, increasing the mistakes and rework. Cats can act out in totally new ways you cannot expect because you never saw it before: shred drapes, vomit, destroy property in devilish ways. Small pets like birds, reptiles, hamsters, you name it, are not safe either. Temperature changes, loud noises and vapors that are only annoying to you are deadly to them.

You may think the workmen will warn you that something they're going to do has been known to kill caged pets like yours. I'm warning you now on behalf of all the workmen who don't see a pet so never think to mention it: your pet can be harmed in ways you cannot imagine if they stay in the house.

Cats are especially prone to poking around, checking things out. If your cat disappears during the work, don't automatically assume the workers left a door open and the pet ran away. They are hard-wired to crawl into new holes, squeeze into openings where they cannot turn themselves to get out, or fit themselves into openings and crevices that are far smaller than you imagine possible. If at the end of the day the cat is missing, first assume it is still in the house somewhere and do your best to rescue it.

Don't ask questions of the estimator for the purpose of determining whether Rocky or Peeps can stay in the house during the work. It's immaterial what their answer is, loud noises nearby can stress out and kill most small critters. No contractor can guarantee no loud noises. Not even a painter. He can't guarantee the aluminum ladder won't slip and hit the tile floor with a deafening noise. Oops to you and me, but Tiger is dead in his cage.

TV, Radio and Computers

The dust of renovation, especially when plaster walls are involved, can wreck electronics. It isn't just the dust settling on them; most sophisticated electronics like computers, TVs, DVD players, routers and cable boxes have small fans to assist with cooling. They will literally pull the dust inside and coat the wires, causing shorts and intermittent issues. Even clocks, radios, and kitchen appliances can have their lifespan shortened by exposure to construction dust. Turning them off during the actual work isn't enough;

the dust is in the air 24-7. For the duration of the renovation, bag your electronics and tape them shut.

During a project, it can't hurt to put brooms, dustpans, paper towels and cleaners in handy sight of the work area. It improves the odds of the workmen doing cleanup during the job or at the end of the day instead of only at the end of the job. If you are living in the house during a several day project, it can be nice to get a little help with that.

On longer projects and those that need several workmen on site for days, it might be worthwhile to ask if they could bring a porta potty and park it in the driveway or behind the house. Doing this also improves the odds that smokers[17] will drop their butts in a localized area around it. Porta potty rental adds to the cost but also makes things less strained for the household and less awkward for the workers. This is feasible only when the weather is decent; it simply isn't nice if it's 20 degrees outside.

I'm Paying YOU to Do This!

This is going to be a few teeny paragraphs, but if you talk to people who lived through major renovations, it's the huge, huge elephant in the living room. There is at least one popular book on hiring contractors that harangues on this aspect for half the book. The thing is this: sometimes your GC doesn't supervise or guide the subcontractors adequately. As the homeowner, you may have to do some of it. It's your house and you want it done right. Your choice is either to rise to the needs of the moment or have things go from bad to worse.

Put aside your resentment and do your best. You're doing it for the good outcome. To make an analogy, this morning on your way to work you are standing at the crosswalk and the little man in the stoplight goes from red to white, meaning you can walk, but you see a truck barreling towards the intersection with no signs of applying brakes. Do you step into the street anyway because you have the right-of-way? No. In life, your choices are a combination of your rights plus what is likely to lead to a good outcome. You know trucks need a certain stopping distance, and this driver doesn't have it. That truck will blow through the intersection even though the taking-turns light says it's your turn. So you do not take your turn. You are a little miffed. After he blows by, you have to hustle to get across, because now the light is blinking, meaning only a few seconds left.

---

[17] Have some tolerance for a localized butt-strewn area. The result of forbidding smoking anywhere on your property will be lengthy absences from work and irritable workers who become more preoccupied as the day wears on. The $6.50 per pack price is motivation enough to quit; making their workday more difficult only hurts you.

Yet nine hours later at home, it doesn't even make a good story. Unless it resulted in a four-car pile-up with lots of blood. But say it didn't, it just inconvenienced you. That's a pretty good outcome. It could have been awful; it would have ended far differently if you insisted upon your rights. But you didn't, and now it's a boring story. That's what you want your home improvement to be: a boring story.

When your choice is to be inconvenienced by extra phone calls, coordinating the work days of two trades or moving materials out of the rain, or on the other hand to insist upon your rights to get the stress-free oversight you paid for, please don't pick the walk-in-front-of-the-truck choice. Choose the good outcome.

Then, at your next neighborhood party, you can tell everyone about your project. "...when my contractor said he couldn't get the electrician to come for a week, I called him and got him to come in two days. I popped in to find he brought only the cheap light switches, the contractor never told him about the ones I asked for, so I told him DO NOT install those, do the ceiling fan first, and ran to Home Depot myself to buy the ones I wanted. Then, the truck dropped off the drywall on pallets in the front yard. I called the contractor and he said just throw some plastic over it and he'll bring it in on Monday, but I said NO-O-O O, it's going to rain cats and dogs tonight so I want guys here before 8 PM to bring it into the garage. And it did, it rained terrible, the front yard was a mud puddle—plastic would not have saved it."

See how boring it is? No one feels sorry for you.

The homeowner who ends up talking to subcontractors six times a week on the phone, coordinating their schedules and answering questions, has a much better outcome than one who sticks with believing it is the GC's job even when the GC is not doing it. It's a natural inclination to hold off, not interfere, and feel discomfited by this strange new mantle of authority.

Most of the time the duties you take on are more in the line of scheduling, reminding, sharing updates with several parties, storing materials in a safe place, and whatever is needed, not really telling them how to do their jobs. You will be their servant in meeting their work needs rather than their master. Often the customer who feels she has already paid for this service jumps into the breach too late, and by then something has to be removed and done over entirely, is damaged by weather, or dozens of other bad outcomes.

Your GC may complain that you shouldn't talk to the workers; the subs may say they prefer to get their orders from only one person. Water off a duck's back. A sad little smile is the proper response to people doing a poor job and getting defensive about it. Just because they prefer something doesn't mean they get it. Don't bend over backwards to 'be nice.' You are not a little leaf buffeted by the winds; you are their paycheck. You are the boss.

When you signed the contract, you believed you were hiring someone to handle this so you wouldn't feel stress. That did not happen. In the end you will feel you pulled this project to completion yourself with your teeth and fingernails. Good job! You know in your heart that it wouldn't be half as nice—heck, it might still be half done today—if you didn't do your very best to help. Absolutely right! Feel good about that.

If to your dismay you find yourself with an incompetently-managed project or your contractor has too many concurrent jobs to do your job well, pick up an oar and start rowing. Where you want to go is the Port of Good Outcome. Resentment is a dead weight you can either keep in the boat or throw overboard. Me, I always throw it overboard.

For the most part, subs are amenable to satisfying the homeowner. If what you're asking wasn't in the original contract, you just thought of it, you need to call the GC and run it by him. Many things are six of one, half a dozen of another if you can catch the sub before he does any work. It's fine to repeat and advise the subs about what was stated in the original contract—they do not see that contract in most cases—but if you're changing scope on the fly, that's something to discuss with the GC.

This is not advocating that you micro-manage, ask them what they're doing every few hours, and the like. If he's putting an outlet where you don't want one, or if that floor doesn't look like the one you picked out, or if they have not covered the hole sufficiently for your tastes, speak up.

Being respectful of contractors doesn't mean hesitating to blurt protests when something bothers you or seems wrong. Sometimes they will explain how it isn't as harmful as you think or is a legitimate expense. The good part about being quick to speak up when something is not making you happy is they soon learn that silence is approval. It's very nice when the workers buy your silence with good work.

Please show unfading self-interest throughout your project. The squeaky wheel gets the grease. A stitch in time saves nine. You get the idea. Always err on the side of saying your thought out loud and having them say "I know! I KNOW!" Then just smile. Maybe they knew, and maybe they didn't, but now they surely do. I take the view that they are just saving face by saying 'I know.' They really forgot but don't want to give you the satisfaction.

Roughly, for every ten sentences you say to the contractor or the boss, one sentence will get to the electrician or other subcontractor. If you don't think your wishes will be adequately served with that ratio, you're right. Make arrangements to talk to the sub directly, in person or by phone. Writing a list, sketch, or other visual aid is great; your GC may pass that along. Put your cell phone number on every page.

On the first day of work, remind workers of the contract points that pertain to this job, plus any warnings and wishes you want them to honor during the job. Do not assume every word you said to the estimator has filtered down. If your estimator is there, cover it again anyway; don't assume he remembers everything.

The odds are good they will not have some of that information. You wanted hard foam insulation? We brought pink batts. It's going to be a half wall? We were told to just tear it out. You're providing the can lights? We brought our own. We can't saw before 8 AM? Nobody told us that. The living room has to be done by 1 PM because furniture is being delivered? Nobody told us that.

In the *Best Practices When Remodeling* chapter there are many small features that will improve your happiness and usefulness—but they won't happen unless you communicate with the boots-on-ground workers doing the actual installation. If you have written these into your contract—and you should, otherwise the contractor will label them extra work and charge you extra—then to ensure they happen you must talk with the subcontractors and workers yourself.

Even if your GC is pretty good, he forgets things. He is juggling a lot of tasks, and one drops now and then. It doesn't hurt to have the workers hear something twice. It hurts a lot to have it done wrong, then either face the cost of tearing it out or living with a less-optimal configuration.

Retaliation

How often do people think if they confronted at the moment they became aware there was a bit of an issue, they might win the battle but lose the war, meaning the quality of the entire job would suffer in retaliation? Sounds theoretically plausible, but I've never seen it happen. Retaliation means a supervisor would say "Guys, I tried to bamboozle the customer for another $300 but she had it listed in the addendum as part of the job, so skip every third nail and scratch up the finish a little." Nope, can't picture it.

Retaliation means the crew lead would say, "Guys, it's outrageous the homeowner didn't know we need to heat the glue before applying so asked us WHAT is that bad smell, so let's take hour-long lunches and track mud into her bathroom." Those things might happen, but not because the boss orders the workers to annoy the customer.

Damaging their own work on purpose leads to more back and forth with the customer, additional cost and repairs. The contractor's workers are like every other worker in the whole world: if they mess up too much or leave unhappy customers in their wake the supervisor will fire them. No company endorses mistreating the customers.

There are passive-aggressive workers out there, or even plain old aggressives, who in disgruntled fashion do bad or incomplete work to hurt their employers. But they don't become bad workers because they are in cahoots with the boss.

The Blurt

An advantage to blurting out an objection as soon as the error, omission or deception registers on your consciousness is that it's likely to catch him off-guard and without a good story.

What happens then is he backpedals in lieu of proceeding. When you catch him pulling a fast one but the act is not fully formed, he can backpedal and claim innocence, with statements like: it was a simple math error, we misspoke, it was a delivery mistake, it's a mix-up with another job, Bob didn't get the word, we just noticed it was the wrong color ourselves, oops the right one is still in the truck, it was inadvertent! It will be fixed right now.

Backpedal or truly a mistake, who knows? What's going to happen now is the inferior materials substitution won't happen, or they've stopped work until the longer screws are brought from the truck, or they attach the flashing before the framing or whatever is the case. They proceed with the remedy under the umbrella that they just caught it too. The result is a higher quality job without knowing if mischief was afoot. This a good outcome. Good enough for me, at any rate.

The biggest advantage to the blurt is when suspicions are unfounded. If there's a good reason for it, the contractor will explain it on the spot which permits me to apologize for my silly outburst and we move on. This happens about 40% of the time; it's not fraud or deception but simply something I didn't know, or my math error. Not a minute of festering suspicions or sinking dread suffered. My trust remains fully intact. Contractor and customer still buddies.

One thing is certain: sloppiness or lack of care will not get better until the customer protests. Retaliation is rare, and fear of it should not work against expressing a concern. In these days of on-line reviews there is no plus and a huge downside to taking retaliation on a customer.

Helping Out

If the work would go smoother with a tool you have or your lending a hand, do it cheerfully. I've even dialed a phone and held it to the ear of the installer who needed to verify something with his office while his hands were too dirty. If you loan a tool it is your responsibility to get the tool in hand before they leave because they will be on autopilot when they throw all the tools into their toolbox at the end of the job.

Punch List

The punch list is a customer-generated list of tasks or actions required before you consider the job done. In the olden days the construction foreman used an actual hole punch to check off completed items, and the name remains to this day.

On short jobs the punch list can be oral, but on longer jobs or those with a several-person crew a written list is better. The punch list often deals with clean-up and appearance.

Sometimes the contractor doesn't come back to finish the punch list. Ask him twice to address the punch list before giving him the option of either doing the items or you will hire someone to do them and deduct the cost from the holdback. If he gives the OK to hire someone else over the phone, write him an email repeating his comments. A good contractor will send you the confirmation email himself.

Punch list items often don't require the same skill level as the main task so employing his staff at their skill level while you engage someone else to do the punch list works for him. Do not, however, pay him the full holdback on his word that he will pay the invoice of the third party you hire. He can subcontract and be invoiced himself without involving you. It's OK to give him the name and price of someone you've lined up already. If he hires them, obtain a lien release from this vendor[18] in addition to all the other suppliers & subcontractors prior to paying the holdback.

Callbacks

If issues arise after completion, then a callback is appropriate and not uncommon. Do not assume you have to fight to get the issue fixed; a contractor budgets for a percentage of callbacks so to him it's just something to schedule. All he needs to know is, does it have to be today or can it be tomorrow or the next day.

Don't try to figure out what's covered by his warranty; it's better to just call, even if more than a year later, to engage him in keeping you a happy customer.

Building Inspector

In construction, many aspects would be shorted, cheated and made flimsy if builders had no oversight or financial incentive to do it right. Weaknesses and inferior materials or methods might take three to fifteen years to fail, long beyond the time you could come after him for remedy. To protect consumers there are State building codes for contractors

---

[18] Addendum A contains copies of several lien releases plus the website to download your state's lien release form.

to follow and the government requires in-process inspections by knowledgeable people, your town's Building Inspector (BI).

Construction permits are issued locally; many have on-line forms and allow payment by credit card. No permit, no BI, no oversight, good luck. The permits give the Building Inspector a heads-up about the type of work that is occurring so he can schedule one or more inspection visits. If he finds something amiss he will often require things get removed or dismantled and the offending part replaced or altered to meet code.

Remove any scrutiny or oversight and your contractor will not do it to code. Taking this view is prudent regardless of the circumstances. Things can be done better than code requires, and in some cases should be; code is merely a minimum requirement.

This is why the Building Inspector is your best friend. Limit yourself to hiring contractors who have a fair to good relationship with the Building Inspector in your town or neighboring towns. When asserting construction is a corrupt arena it might imply it extends to defining a 'good relationship' with the Building Inspector, if you catch my drift. One has to hang their hat somewhere, and I hang mine here: *Nothing positive or cost effective comes from an antagonistic relationship between your contractor plus any one of his subcontractors, and your town's Building Inspector.*

Someone might fill your ear with the flaws and shortcomings of your BI, but that is irrelevant. You need a contractor who has figured out how to work agreeably with him or else it will cost you delay, frustration, tear-outs, and most of all, lots more money.

The BI is like an Umpire in baseball; some players use their experience to get away with skirting the rules when he isn't looking and quarrel with his interpretation when he sees it. Other players never have a problem with an umpire. Since the rules are clear-cut and there isn't a game to win or lose, it seems plausible a contractor could just comply with code to eliminate all hassle in his life. Try to find that guy. For the rest, there's $5.50 to gain so the temptation to use the cheaper wire, grade of wood, or you-name-it is a siren always singing.

No matter what your opinion is of the Umpire, if baseball were played without them, absolutely every pro, no matter how nice and friendly they are, would play fast and loose with the rules. Without any risk of being thrown out of the game or fined, players would ram the second baseman as they approached. After hitting the ball, the tossed aside bat would accidentally bounce off the kneecap of the catcher. They don't do these things in baseball because they're watched and there are serious ramifications.

Your contractor knows ways to 'win,' as in make more money today, by seriously wounding your future financial health. Since your BI can't be there every day, simply mention him frequently and praise his sharp eyes.

Don't be fooled; don't accept it lock, stock and barrel when the contractor talks about the BI as if he's a spoiler, a wild card, someone who strolls in and ruins everything. The BI is your best friend.

A contractor may attempt to enlist the help of the homeowner to evade compliance with code, explaining it in some way that sounds like it's a good deal or makes it seem as if the code is unreasonable or outmoded. This is so seldom the case that even if you have no means to articulate an objection and can't imagine what the counter-argument would be, your answer should be that you want it to be code or better.

The situation might not be between code and below code, but better than code and code. Say a certain feature could be VV or w. The Architect calls out for feature VV on paper, and later your contractor finds some moment when your concentration is divided to mention that VV is not to code and he would like to do it to code, which is w. At the moment you cannot evaluate VV vs. w so you say OK. The Architect's feature could cost three times as much; in getting you to authorize stepping back to merely code, the contractor, who quoted the more expensive thing, gets to pocket the difference. Just because a contractor says it's code doesn't mean it's better.

Code is a minimum.

Put another way, code defines the very edge of bad. Beyond that is illegal. For things like steps, windows, floors, insulation, lighting and so on, no one should pay big money just to border on the edge of unacceptable.

Many times I specified features better than code and contractors have foot-dragged on virtually every one. It would be unlikely I am the only one to suffer push-back. A house that only meets code would be boring. Upgrades from code add style, beauty and convenience.

Do not allow your contractor to work in a manner that makes it more difficult for the BI to see his work. This means all the electrical lines and plumbing lines should be visible and inspected by the BI, and only then may the drywall be installed. If your contractor drywalls over newly installed wiring or plumbing before the inspector takes a look, the BI is perfectly within his rights to demand it be torn down. For this reason it's not just personal preference, it's a huge waste of money to take that risk.

Even if the GC doesn't forward that cost to you, don't think you dodged the bullet. One of your largest risks is that he will go out of business during your project. Bone-headed sequence mistakes like this contribute to him not making any money. Save him from himself, at least until your little piece of the world is done.

Getting Your Money Back

Most lawsuits stem from unmet expectations, reasonable or unreasonable. What actually happened can be very different when the shading of expectations is removed.

A lawyer once proved it with an example like this: A manager walks up to one employee and says: "You've done a great job. Here's a check for $200." He feels appreciated, right?

A practical joker sees this, so scoots up to the next guy and says "The company is giving away $500 checks to the good workers. Look, here comes your boss." The boss hands the guy a $200 check and says "Thanks for your good work." Feeling unappreciated, he goes home deflated and disappointed.

What happened just before, causing an unmet expectation, can lead to severe dissatisfaction *with no change in the facts*. Finding out early in the process that an expectation is not reasonable robs it of its power to cause unhappiness. In the case above, if a fourth person overheard the joker and blurted out "He's kidding, all the checks are for $200," it would have eliminated the negative reaction.

Some lawsuits germinate not with one big thing but several little pebble-in-the-shoe issues and then one impertinent poke in the side pushing it over the edge. In court, the poke in the side incident stands by its lonesome, making the plaintiff look touchy and unreasonable. These are nearly always won by the defendant.

For dollar amounts within small claims court jurisdiction (it varies from $2,000 to $10,000 by state), you can speak on your own behalf before a judge and receive a judgment. This chapter assumes all situations are small claims court material.

To win money, you have to prove money losses or damages. 'Prove' and 'money' are the operative words. You can only be reimbursed. Pain and suffering or punitive amounts are granted only on top of actual damages. When the actual damages are dinky, forget about getting huge amounts because the other guy deserves to be punished. That only happens if the other guy pisses off the judge, while you are sweet and nice. If medical insurance covered all except $88 and you missed no days of work, then $88 is all you may win.

Remember to bring witnesses. Only police reports are allowed as absentee statements. Bring receipts, bank statements, and photos. Many people get so caught up in their narration of the situation they forget that in a court with only money, not guilt, at stake, only the money trail matters, not the he-said, she-said. If your case is that you paid him $4,000 and got only $2,300 worth of work, if you don't bring solid evidence that you paid him $4,000, you will lose. The other guy won't need to say a word.

If you already know this then bear with me, but winning in court does not mean you get the money. A judgment is not a settlement. The money comes from the same place it did before, from the defendant. If he or she had no money before, now with court costs and time off work they have even less. The effectiveness of measures forcing the defendant to cough up the money owed can vary widely from state to state.

In many states there are so few ways to force compliance that some winners in court don't see any money. If the person you are suing is a student, self-employed, unemployed, lives out of state, is a dependent of someone else, is in Chapter 11 (bankruptcy) proceedings, doesn't own real estate, has social security as their sole income or is employed but their boss won't cooperate with a garnishment, then the usual compliance measures won't be available. You can't begin any measures at all until the state's allowed time to pay is up, which could be 30 to 90 days.

Perhaps the defendant pays within the allowed time so taking him to court was worth your while. If not, every next step to get your money will cost you, the winner, more money. None of this can be reimbursed since the case has been settled for a fixed amount. Some states have stern language on their websites vowing they would throw the non-paying defaulter into jail for contempt of court. To anyone thinking of defaulting on a judgment; pay or else suffer the consequences.

Grievance?

When you have a grievance, think hard if your expectations are shading the actual incident. To point out the obvious, every person who fires up enough head of steam to take another person to small claims court felt at one time they were indisputably wronged and ill-treated by the other party. Yet by the time the facts are aired in court, few cases are clear-cut.

That the defendant wins even 10% of the time would be astounding if the plaintiff was always wronged. Yet the defendant wins a much higher percentage than that, meaning the judge decides the plaintiff can prove no misdeed worthy of a monetary settlement. Even when the plaintiff does win, half the time it's a squeaker of a judgment call or a partial victory.

A project can end up with a lawsuit when the customer doesn't speak up or word it clearly enough for the contractor to pick up on it. For instance, a customer might assert "I told him three times it was too loud." but never asked for the noise to stop immediately.

Another customer has a guest with a toddler coming over that evening so wants to convey that the hole in the floor can't merely be surrounded with three orange cones but must be sturdily covered if not completely closed by 5 PM today. Stepping into the work area, the customer points a finger and says "Don't leave that open like that when you go home." The contractor, glancing up for only a split-second, thinks he's talking about the open window, not the hole, so replies "Yes, I'll take care of it." When it's important, it isn't enough to have two sentences filled with vague words; make it a longer dialog describing the details and repeat your understanding of his response.

In the case above, the homeowner would reply, "So you agree that by 5 PM you will have the hole in the floor securely patched so a small child could not fall through it?"

Not Actionable

Sometimes a lot of dust, dirt, noise, delay, paperwork, lawn damage, oily residue or bad smell is part of the job, meaning absolutely unavoidable. Complaining to the contractor about it is a waste of breath.

It is reasonable to ask forward-looking resolution-oriented questions like "How long will the bad smell last?" or "Could we not use the sander before 8:00 in the morning?" or "What is the plan for fixing the lawn?" and these should be answered honestly, even if not always to your liking.

In construction and installations, there are a lot of things that need tweaking in. An expectation that every install or build is perfect on the first try is too high of a bar. What you really want is prompt tweaking in when you place a callback. Maybe even three or four tweaking-ins before it's done. Don't get upset or full of anxiety if you need to call the shop because a leak appeared, you spotted a gap, something fell off, something doesn't work, or you name it; be confident that they are expecting at least one callback.

The cost of the callback and later warranty work is included in the quoted price. It is no charge to you.

Be in good humor and help them get in and fix it as soon as possible. Judge the job on how well you are pleased with it after the callback, and not ding them for simply needing to come back to get it completely perfect. After all, in the end you have a wonderful job that was done by them for $xx amount. Make the outcome be the final word.

Individual tradesmen can be as honest and hardworking as the day is long. When dealing with your contractor's workers, keep in mind that everyone involved is not necessarily in cahoots with the worst of the bunch. They may frown upon it. If you find yourself at loggerheads with a certain person, even a high-ranking person, a good ally may be another employee at the same company. Try to find and enlist that person, be it the receptionist, the accountant, a Supervisor, another Leadman or whoever, to speak on your behalf. Don't immediately fly to blaming the company when it's just one individual with customer relations issues or poor planning skills.

Don't Go to Court

Here are several better approaches with far greater odds of providing satisfaction than going to court. These should be tried and exhausted before resorting to small claims court. Court can be a more unpleasant experience for both parties than a first timer imagines. Baggage scanners, pat downs at the door, long waits surrounded by unsavory people. In many cities just finding parking near the courthouse is difficult. Leave children at home because a 'take your child to small claims court day' could be a scary and miserable experience for them.

Before Going to Court:

1. Threaten to damage his future business to scare him. Tell him you're going to write a tell-all bad review on Angie's List, Yelp, Yahoo or other review site. Let him stew on that for a day, then tell him you're thinking of making a complaint to the Better Business Bureau (BBB.org). If that still doesn't make him budge, check his business card and website for professional organization membership and say you're thinking of filing a complaint with them too. Many people use on-line reviews to pick contractors, so these measures will have significant repercussions for his business. Here's the kicker: all the effectiveness is in giving the business a chance to prevent it. There's no clout to wield after you've done it. So mention it at least twice before actually doing it.

   Alternately, you could try the 'carrot' approach instead of the stick. Tell him if he fixes this matter to your satisfaction right now, you'll write a good review as if it was his course of action all along. Give him a window of time, a few hours or a day, after which the opportunity for a good review disappears.

2. If he still won't budge, then make the on-line reports to the BBB, his professional organization and Angie's List. Now give them time to ask their questions and put their little muscle to work on your behalf. A good percentage of the time that's sufficient. Still, it's possible he wiggles out of it.

3. Have a lawyer write a very intimidating letter on official stationery. Asking a lawyer to do this usually costs less than any other legal action, something on the line of $50 to $150 because they have lots of this sort of letter on their computers and they borrow whole paragraphs and change only a few words to suit your situation. I have seen a few of these letters and they are impressively scary.

4. Inform him you are prepared to file a complaint with the State Attorney General's office. If that doesn't soften up his hard stance, do it. Most states, if not all, have a good AG website. you can find it by searching (state name) Attorney General. Many states provide mediators for resolving disputes with small businesses, car repair, home buying or landlords. The possibility of a complaint with the state bothers a businessman more than going to small claims court.

5. If you used a charge card for some or all of it, tell him you will contest it with the charge card company. Generally, you have 60 days from the date of the charge to contest it. The reason has to be actual non-service, non-delivery of what you ordered/delivery of something cheaper, or a charge you didn't sign for or agree to. More details are in the *Credit Card Court* sub-chapter.

All the effectiveness is in warning before doing. Give him time to avert it. If you truly want a good outcome and not cruel vengeance, let him know what's coming and giving him 'one last chance' a few times. This advice applies only to legal, above-board situations when you wear the white hat. If anything tax evasive, permit-forgetting, shady, or under-the-table was involved, most likely the other guy will draw that sword and you'll bleed more than he will.

There are times when you want to ask for some of your money back in mid-job or even after the job is done, if that's when you find out you were definitely overcharged. I get money back after I've paid all the time. After landing upon an advertised special or coupon, I go back to collect. Finding out they'll match a competitor's price, ditto. With businesses dependent upon good word of mouth, a deal isn't over until the customer is happy.

What if the job was completed two weeks ago, you paid in full, but when talking to co-workers you find out it should have cost much less? I would phone them with a head of steam and curse at them, you lying, cheating SOB . . . but that's my first thought. My second thought is a much better idea: pull yourself together, prepare your time and materials evidence, jot a little list of bullet points, think about what would make you happy, and then make the call.

First, get the right person on the phone before you start spewing your grievance. You want all the majesty of your indignation to spray onto the right person, not the clerk. Tell the phone answerer that you have a concern about your bill, and wish to speak to the owner. Maybe you get him, or maybe you get Number Two or the accountant. With luck they route you to a person good with customers. Wanting to speak to the owner about a bill gets their head in the right place: worried.

Whatever approach you take, always behave as if a reason you couldn't imagine right now might make the charges fair. That happens sometimes. I'm not saying don't sound upset; I'm saying be able to turn off the anger if it becomes unsupported, or conversely, if they comply with your request. If you aren't the kind of person who can shut it off like closing a door and sound contrite or grateful from that moment forward, then you must do the whole call in a reasonable voice and kindly tone. Those are your two choices.

Why? Because the goal is for this company to return you to happy customer status by charging you the correct amount. If there's no possibility of you sounding happy or being happy, what is their motivation? Why would they give the money back if you're going to lambast them on Facebook anyway?

After they access your record, proceed to share your new details, share how you know you were overcharged and what you think the amount of the overcharge is . . and then ask for an amount less than that. The logic is that it's proper for you take a little hit for not being more on the ball. For example, if they charged you $3,000 to replace the front door and you find out $1,500 is a fair price, tell them you would like $1,200 back, or else . .

What is totally out of the question is taking him to court. Won't work. You two have a signed contract, end of story. The only tools you have to work with are threats to his future business and being a pain in the neck.

After saying you contest the overpayment and letting him know the amount you wish returned, ask him to call you back in two hours with his decision. Not rushing him greatly improves the odds of getting what you want, because people's second thoughts are always better than their first thoughts, and his first thought is 'go to hell.' He was probably just about to hang up on you anyway.

Cooler heads may prevail. When you call him back three hours later, move on to mentioning you are writing a draft of a bad review revealing the overcharge, or will contest it with the credit card company if you charged it. If you feel like it, sweeten the deal by saying you are willing to write a good review (if you are happy with the work) if they modify the price to match the work. You'll basically give him a mulligan on the first price and write the review as if he charged you the second price all along.

My only other advice is to squeeze all the juice out of every 'or else' before tacking on the next. Let your increasingly bad-sounding proposals pile up until they tip the scales. After phone call five or six with no budge, implement. But odds are the two of you will be doing back-and-forths about the dollar figure before that, until you agree to something. Behave ethically and wear the white hat; it's bad karma to do anything less.

Debit Cards in Real Life

Pay bills, invoices, down payments, good faith money, milestone payments, and holdbacks with a credit card or a check. Never use a debit card or cash. A money order or cashier's check are a shade better than a debit card, yet those leave you with less-accessible cashing date and signature information and are *weaker in a court of law* than a personal check.

Debit cards often have a MasterCard or VISA logo on them, leading people to believe they are somehow like a charge card. Not true. A debit card is merely a fist-sized hole into your bank account with a sheet of nose-wipe tissue covering it. Money doesn't exactly fall out of it, but it takes a miniscule effort to grab it out.

If anyone does steal from you using your debit card, not only is the bank uninterested, the police are even more indifferent, requiring you to find the culprit yourself and prove it. Even then, for some inexplicable reason it's seldom prosecuted as a felony.

Banks love debit cards because users average over $200 per year in fines. You must specifically request an ATM card because the bank will automatically issue a debit card. Ask for the ATM card and cut up the debit card.

Shockingly, several people have told me straight-faced that they feel debit cards are safer than in the past because the bank will freeze the checking account after some untypical purchase. We all make untypical purchases now and then. Meaning our other checks bounce at random intervals, causing huge inconvenience to the card-holder. The bank doesn't just close the barn door after the horse is gone, it smashes the door on the owner's face.

The analogy is a perfect one, because with a debit card the bank does not get the horse back for you and doesn't care that their knee-jerk too-late action inflicted more harm. With a credit card, the horse magically appears back in the barn and you are uninjured. If anyone makes unauthorized charges or withdrawals from your charge card, the most you are liable for is $50, and many cards will waive even that if you report it within three days.

An ATM card is safer because all users pose for the cameras at ATM machines, there's a very low withdrawal maximum, and users can check the balance before withdrawing, preventing fines.

Credit Card Court

If you paid by credit card, contesting the payment with the credit card company is absolutely the most perfect, wonderful method in the whole wide world to get your money back. It's easy, free, and quick. When you win, your money appears instantly. If you lose, no court costs. No paying for parking, no taking off work; simply sit in front of your computer or whip out your phone during a TV commercial break and do it.

Paying with a credit card puts all the power and clout of the huge credit card company on your side when you have a dispute with someone you paid for goods or services. Even if the charge posted fifty days ago and you've paid the card in full twice since then, you can dispute the charge and get your money back if some deception or non-delivery occurred.

Never before in the history of mankind has there been such a powerful entity enforcing ethical behavior in every type of business. It is a superpower we should not discard lightly by using checks or cash.

To contest a charge, go to your account on their website, find the charge, and click to contest it. The actual steps vary by site; some have you fill out an online form with the transaction ID, others have a clickable link right below the transaction line. Look for the FAQ; it will always tell you how to contest a charge.

The differences between contesting the charge with the credit card company vs. using small claims court to get your money back are so profound and amazing that you will wonder why no one told you this before.

When you contest a charge in 'credit card court' as I affectionately call it, immediately, the disputed charges are frozen, meaning no interest accrues, until it is resolved. Card policies could vary, so if you are carrying a balance, ask the customer service rep. If you already paid the full balance owed before or during these proceedings, it has no impact on your case, meaning it doesn't imply you think you won't win.

When you take it to small claims court, interest keeps accruing, along with fines if you don't pay your bill.[19]

---

[19] In court, most people forget to include the interest payments when the judge asks them the amount owed, and forget to include interest payments that will accrue over the next two months because it might be 30 days before you see a check. It's a huge mistake to list several items as if the judge will add it in his or her head. When the judge says "How much," what pops out of your mouth should be the absolute gross total, then silence. Don't forget to include the parking fee that morning. If the judge wants a breakdown of costs and wants to tickle some of them out, fine. But if you start out itemizing, the judge is likely to cut you off, then simply rule for only the initial cost—no interest, fines or unpaid time off work.

Credit card court leans in your favor. Yes, I'm saying the judge is your buddy. Put another way, vendors are held to a higher standard than card users, which is the less sensational way of saying the judge is your buddy. In small claims court, the judge is impartial.

Credit card court looks closely at past bad acts for patterns. It may take as few as three contested charges for MasterCard or VISA to pull credit card services from that company. You do not stand alone. If this company is building a history of complaints, those act as a multiplier. In small claims court, past bad acts are not admissible.

The credit card company will send you paperwork immediately. Watch for it and endeavor to fill it out and return it the very next day. They move fast and will phone you if things go a few days with no response. Remember, you have mobilized your buddy now and this business's lifeblood hangs in the balance so they want to clear it up one way or another as soon as possible. Small claims court takes months and memory gets foggy, harming your case.

The credit card company will communicate what the seller says and will arbitrate. You do not have to speak to the seller if you don't want to. In small claims court you have to share a waiting room and then sit ten feet away from him in court, which is upsetting. The credit card company may ask you to send photos, documentation and whatnot. The defendant is asked for similar evidence. He cannot win by sitting silent, which can happen in small claims court if you don't have sufficient support for your accusations.

If the card company decides in your favor, your money comes back immediately. Since new customers are using their charge cards with that business every week, it's no skin off the credit card's nose to set aside part of it and give it to you. The seller gets dinged and you are made whole. In small claims court, at the very best he sends a check 25 days from now; at worst, you have to pay more money to initiate collection proceedings that might snag some money months later.

A business often leaps to settling with you directly rather than have the credit card court decide against him (to avoid the multiplier effect in future disputes). That's fine. Online or in the paperwork they send you there will be a checkbox next to a statement saying that it has been resolved to your satisfaction. Check the box, sign the form, and the credit card company stops mobilizing on your behalf.

Complaints and Warranty Work

Things go awry: something is broken, the button doesn't work, issues are discovered. You want to phone the business and yell at whoever picks up. There's an overpowering urge to give an earful on how bad this is for you, how much you're inconvenienced.

Deciding what you want the guy on the other end of the line to do about it is the place to start.

What we really want is for him to make it all better. Can he do that? Sometimes. But usually he can't wave the magic wand and made all the disappointment and bad repercussions go away. On occasion there will be something he can do or advise you to do that will quite unexpectedly make you whole again.

One thing that might help you bite back your peevishness is to realize being angry stems from being out of your element and not in control of the situation. The guy you're calling, however, IS in his element and has some control; therefore it stands to reason he might very well have tools in his toolbox for this type of situation that you don't have or aren't even aware exist. Enlisting his help to get to your desired outcome has good odds of success. Keep your good nature, at least at the beginning of the call, so he has an opportunity to be a hero.

When a business has let you down in one way or another, before calling regarding the issue, think a bit about what you would like.

Here are the basic bones of the choices:
- Money
- Replacement
- Service, promptly please
- Future price breaks or additional items
- Free stuff
- Or some combination

Before picking up the phone, decide what would mollify you and what would make you happy. What would make you not pursue it any further, and what would keep you as a satisfied, happy customer. Be prepared to give him the choice.

Example: Several years ago I had an issue with a printer-scanner-fax from Hewlett-Packard. It jammed a lot. After about eleven months of struggling with it, the readout went odd, with a few words of programming language showing. So I phoned HP. After reaching the technician, we went through the obligatory trouble-shooting and reset procedure over the phone for about an hour. We actually got it back to working well. But a week later it happened again.

They wanted to do the reset again. That did not appeal to me. There was no reason to think it would last this time and my warranty was almost up. They wanted me to box it up and send back for repairs at their expense. I was unhappy to spend a few weeks without a printer. They offered to send me a loaner.

These were all appropriate responses to my objections, but the piecemeal approach was becoming complex. Visions of dumpster diving for a box and using my lunch hour to troop to the Post Office went through my head.

I offered what would make me happy is a new printer. Those were my exact words, 'would make me happy.' After that request I was put on hold for several minutes. My technical rep returned and said this model is no longer available, so a replacement wasn't possible.

An equivalent or better model would be acceptable, I replied. Again I was put on hold so he could troop back to Mr. Packard's office to ask for the OK. That was the mental picture I amused myself with while playing Solitaire on the computer. My rep returned with the good news that they would send me a model 3310, which is their latest printer-scanner-fax of the same size with even more features. I said "That would make me very happy."

They asked me to put the old one in the box the new one came in, put it on the porch and someone would pick it up the next day. No problem. It did not escape my notice that my complaint about trooping to the Post Office was also solved.

When they can grant either your mollified customer or happy customer choice and they chose the happy-customer choice, thank them with a happy tone in your voice. There is no more reason to punish them.

The whole point of dealing with complaints and issues is to restore you to happy customer status, so there's no point in considering it an anomaly in the function of the universe. The point of being a business is to have many happy customers who are satisfied.

Phoning the vendor while holding the mental image that disappointment is inevitable is a self-fulfilling prophesy. Expect fair play on warranty issues and you quadruple the chances of getting it.

Assumptions about the other guy's costs and operations can make customers hesitate to say what would make them happy. For instance, if the leadtime on the first job was four weeks, it might appear there's nothing he can do in three days. Put aside assumptions and describe what you would like to see happen.

Leadtime is a tool for scheduling company assets plus enabling purchasing quantity breaks. It is often unrelated to the actual length of time the work takes. Don't assume before you call that he can't fulfill your first choice to resolve the problem in a short time. Share as much information as you can; perhaps he can connect some of your dots with his dots and come up with a better solution.

Having an idea of a good outcome may not come easy in some cases, but don't pick up the phone until you have at least one option. Some are toughies. For instance, if equipment parked in front of the garage at 5 AM caused you to miss the plane and your sister's wedding, he can't turn back the clock. Yet covering only the $75 flight change charge seems inadequate. If they left a tool in the hallway and you stubbed and broke a toe on it while going to the bathroom Saturday morning, but your insurance covered the medical costs and no work will be missed, there's no court case there because little value is placed on that level of pain and suffering.

If your case is a lifestyle ding and you come up with something that you feel is adequate compensation, you'll have to sell it, not give ultimatums. It isn't small claims court material. Appeal to his sense of fairness.

Most situations will be easier, with a clear remedy. Write down key words or lists on a piece of paper, and perhaps the opening line verbatim so you don't get tongue tied. Then make the call.

Start out with a warning that you are calling with a request for service or an issue with the job or product. At that point the phone answerer may interrupt to ask if you would like to talk to another person. That is excellent. At small to medium companies the phone answerer could be anyone in the office, meaning the bookkeeper, office manager, or merely the newest person. It is not a sign of poor management if the phone answerer has neither the clout nor the understanding to give you what you want.

Whether you're passed to the boss, service technician or scheduler, tell each person what you think the issue is in the same mild tone of voice. Perhaps the situation requires explaining what should have happened or what you'd like to have happen. Finish up with an explanation of how it affected you.

Using this system of handling warranty calls means every now and then it results in more than expected. This occurred to me during a periodic equipment maintenance visit on a ship docked in Hawaii. An electronic tool needed a part, and repairs on the ship would come to a halt by the end of the day without it. With time of the essence, overnighting the part from the mainland seemed to be the only choice.

The East-coast manufacturer informed me that they did not keep it in stock; the tech rep could only provide the name of the small company that manufactured it. Googling the small company revealed it was in Texas. I called the number and after a few menus, reached a sales rep. While she was looking it up on her computer to check whether it was in stock, I mentioned that getting it as soon as possible was important. If it could be overnighted today, I would be so grateful. Since it was past 2 PM in Texas it seemed hopeless, but I asked anyway. After finding over twenty in stock, she asked me to hold.

The hold lasted almost half an hour. Upon her return, she instantly apologized for taking so long, explaining she had been hunting for sales outlets in Hawaii and making arrangements. She found the part in stock at a

Best Buy only a few miles away from me; she had phoned them to arrange selling me the part. As it happened a co-worker was out on another errand; we reached him on his cell when he was about 30 seconds from passing that exit. The part was literally in my hands twenty minutes later.

The unexpected rule of thumb about warranty and service calls is: the longer you are kept on hold, *the better the outcome*. Never hang up no matter how long the hold is. If the call is dropped, call back again immediately. This rule of thumb has been true so often for me that my spirits tend to soar when the wait goes over fifteen minutes. In the case above it started to feel like the rule of thumb wouldn't pan out, and then it did, with fireworks and marching band.

Major Renovations & Additions

Before you begin to stick your toe in the waters of a major project, check the local zoning laws. Talk to your building inspector about the laws pertaining to your neighborhood. These might affect how close you can come to the edges of your property, limit square footage, limit the purpose of the addition, and un-grandfather you in some aspect you will never be able to guess. "Grandfathered" means it was in place prior to the exact date the zoning law became effective. Thus, you are not compelled to do anything because your home was legal and remains so. When you modify part of the house, now you may be required to comply with current zoning to pass building inspections.

In many areas the homeowner's association (HOA) has an additional layer of rules, whose obedience is compelled in the fine print on your deed. Condo associations also have rules that affect what individual homeowners can do to the property, inside and out.

If you overstep any of these rules, the law is on the other guy's side, the side that says you have to tear it all out. The scope and breadth of these rules are impossible to predict or cover in one book, and have no doubt some of them are ridiculous, unjustifiable, and nit-picky to an extreme. So look before you spend any money.

In some communities the structure can come within three feet of the property line, in others ten feet, in others thirty feet. The rules can prohibit non-structure changes such as installing a fence, a shed, a treehouse, a paved patio, even planting certain kinds of trees and shrubs. The last thing you want is to pull a permit, get halfway done, and have the inspector's visit end with a big, fat, never-changing NO.

TV shows about the exploits of home flippers and remodelers show them wasting tens of thousands of dollars on aborted plans—and one would think they have the staff to look into zoning laws and HOA rules. Even "This Old House" had to change a project in the middle of filming! If I recall correctly, it involved raising the roof to create a third floor; a great idea, but not when zoning doesn't allow three-story buildings in that area.

In short, even though you own the house, you aren't free to do what you like. Before you hire an architect or designer, before you call a contractor to the property, be sure you have sleuthed out the maximum parameters and outer appearance rules yourself, or hire someone specifically to do it. No one else will naturally do this for you, and the further along the project is, the more everyone assumes it's already been done. There are 10,000 stories out there about additions that went several inches over the allowed footprint so had to be removed or chopped.

When you find out ahead of time that zoning doesn't permit it, it's possible to request a zoning waiver or change the zoning rule. If your situation is reasonable, it's likely you'll get it. It could take months, but it is the lesser of the evils. Doing it afterwards won't have as good a result, because zoning laws are usually not retroactive, which means you still built under the old laws so are still in violation. Besides, you have pissed off the city fathers with your insubordination.

Zoning and HOA rules are cases where the old saying, "It's better to ask forgiveness than to ask permission" does not apply. Don't test me on this; trying to sneak a violation through, hoping no one notices, is too risky when your financial ruin is at stake.

Going forward with an addition, major renovation, or even getting a whole home built, you will have a lengthy relationship with your contractor. There are contractors and General Contractors, or GC; the latter serve as the boss of a temporary conglomeration of several specialty contractors. The GC may do part of the work with his own crews, or the GC may have no workers on-site, but only hire, manage and coordinate all the various trades needed to complete the job. When you hire a general contractor you pay only him, and he hires and pays all the other guys. He provides the warranty so you don't have to worry about who made the mistake. That's the theory, at any rate.

A homeowner may serve as his own GC. Since it amounts to hiring several individual contractors, setting their work schedules, and shopping at the hardware store and lumber yard, it's conceivable you could give it a shot. Every state has different rules, and only a few have test-passing and experience requirements for GC. Being licensed means he hit some bases, but when hiring a GC you still need to follow the multiple quote process for hiring contractors.

Visit this website for information on the rules in your state:

http://generalcontractorlicenseguide.com/

The more building experience a person has the better GC he is. If you haven't worked in any of the usual trades—drywalling, cementwork, plumbing, electrician, landscaping, roofing, framing, etc.—in a paid capacity, only dabbled in your basement, you haven't got the chops to do this job without many regrets at the end. You have no idea of the extent to which a major project can go off the rails even though you take every precaution within your ability to predict.

To be clear, becoming a first-timer GC on a home build, addition, kitchen remodel or major floorspace redesign, usually with the notion of 'saving money,' is the path to developing facial ticks, insomnia, and losing the ability to smile for two years, not to mention will put the worst kind of strain on your marriage. Woe be if you are engaged or dating, because that's over now. Remember that little quip about never hanging wallpaper with your fiancée or the marriage is off? This is equal to twenty rooms of wallpaper.

It bears repeating, you want to hire experience when you hire a contractor. You never want to be a first-timer; it will not become your dream house, it will be a nightmare.

You want him to be past al-l-l-l the beginner's mistakes on five other people's houses. Or thirty.

What kinds of things can happen to an inexperienced GC? To throw out just one thing, an inexperienced GC will incorrectly predict lead time on materials delivery so will incur waiting and delay expenses. In some cases delay exposes the project to weather damage, which must be repaired . . . and so on. There's no way to fast forward twenty years of experience into three months. The time and cost estimates of an experienced GC are more reliable and accurate. He's been this way before so can see around corners.

As in all careers, there are people with ten years' experience and people with one year of experience over and over ten times. Ask for references and talk to the people who hired this contractor in the near past. If he can't provide recent references, say within the past 15 months and especially something he just finished, find a different guy. Even if he was good two years ago, his last four customers could be experiencing legal action from unpaid subcontractors.

When you talk to the owners of his past jobs, you will quickly see if he ordered materials efficiently, kept the job progressing, honored customer wishes, and kept in-process damage to a minimum. He is cherry-picking the jobs that ended well, in his estimation, to use as references, so if they had problems, you cannot expect better.

It may be the case you'll never find a 'perfect' GC in your area. Actually, most likely you won't. But each has different strengths, so choose one and resolve to help him or her as much as you can.

Building and construction trades are prone to mismanagement. Home improvement/construction issues are the number three cause of consumer complaints, just behind auto-related issues and credit card/loan/fraud issues. Considering the average person in a typical year may have eight auto-related interactions and one hundred credit card or loan interactions yet only one construction interaction every year, being high on the list with those other businesses indicates a complaint probability ten to one hundred times greater per interaction.

Re-read the chapter *Watching the Work*. The larger the project the more true it is: if you're spending big money, you have to watch. Search for Youtube videos on various segments of your project. For a few days, looking without understanding can suffice to keep them working well. If this project is longer than several days, they'll figure out you can't discern they skipped caulking around the window or didn't put up a vapor barrier, which are two typical cheap-out maneuvers. After that your staring at them won't prevent anything.

If you get enough quotes to avoid both the swindlers and the opportunists, mind the contract so it includes the verbals, watch the work sufficiently, have permits that provide BI inspections, and keep payments lagging behind work, either you will avoid reasons to complain to the state or if he's incorrigible you can fire him and get someone else to complete the job without taking a big financial hit.

Subcontracting

Often a contractor will subcontract part of the job to a tradesman he's worked with in the past. This usually is invisible to you and works out well. Although you don't get to pick the subcontractor, the odds are good they both share the same quality level and your main contractor trusts that he will not screw up.

The pluses for you are that you pay only one contractor and that contractor takes responsibility for the whole job, so if anything goes amiss you don't have to deal with finger pointing over who caused it.

Hiring one trade and expecting him to subcontract to more than three other trades, making him a de facto General Contractor, is another matter. If you're lucky he will reject it, but if you are unlucky, he needs the money so he'll do it. Then you have a first timer GC who will not give it the same scrutiny and care that you would. If things go badly and he's into the red, it can be cheaper for him to walk away from your job. You'll be left with a half-done project and not a single option that has you paying less than double the original cost unless you can get a show on HGTV to feature you as a project.

Project Scope & Change Notices

No matter how much research and work you put into it before contract-signing, more thought results in better ideas. This means after the work has begun, you will have changes. Don't feel bad about this; only really dumb people can turn off their brains for weeks on end.

Major Renovations & Additions

As mentioned in the *Contracts* chapter, the contract should say something about the forms and methods for presenting and approving change notices. If a pre-printed change notice form isn't available, grab a piece of any old paper, write down the change, and both sign. Throw it on the copier; done.

Whether it's your contractor saying he's moving the door two feet from the plan's location for some reason that saves him work, or you asking for bigger windows, or he's not going to buy the lumber from the lumberyard mentioned in the contract, write it down, write it down, write it down, and both sign. Don't limit this to changes that affect the price; it should be done for all changes from contract.

It's a good practice for your main contract to state that when a change entails a price change, the plus or minus dollar amount must be noted on the change notice or else there is none. This prevents him from saying OK to bigger windows[20] but at the end of the job saying you owe him another $900 for the cost difference plus restocking fee for the original windows.

Your contractor can keep a note pad and write all the changes down, for you to sign at the end of each day or as they happen. Or he can email, and you reply with an approval email, if that's what you want to do. Or you write down what he said over the phone, and have him sign it later. The reason to write down changes, with signature or approval emails, is because when you wink at your own contract, meaning verbally discount this or that or give a verbal OK to not obey, later on when something you really want doesn't happen, he will say you verbally approved it, along with twelve other big changes you verbally approved. You don't have a leg to stand on. You cannot prove you didn't do it this time.

When change orders involve new features, say a bay window in the breakfast nook or a tile fireplace all the way to the ceiling instead of only to the mantle, you should expect the cost to be the additional labor cost plus materials, with a mark-up of 15% to 25%. Asking to see receipts for materials might be in order if the contractor provides an unusually high estimate.

Contractors tend to go a bit opportunist on contract additions or embellishments. You should see an estimate prior to start and have all price discussions before that additional work begins. You will have to be timely to

---

[20] He knew from the minute you said the words that it would entail a higher cost, but at the time either he wasn't in the mood for price haggling or didn't have time to collect data and do the math, so wasn't going to pull a number out of a hat. He saw less downside to just dinging you for the cost later. This practice of mentioning change notice costs only at the end of the job makes for unhappy customers who don't make good references for his next job. It's a bone-headed move on the contractor's part. Helping him avoid bone-headed moves keeps him solvent and who knows, you might even become a good reference for his next job.

a compulsive degree, because the alternative is having workers standing around doing nothing, a cost the builder should actually bounce back to you because customer-initiated downtime was not in his initial price.

Depending how busy the contractor is, change notices might be welcome extra work or a huge inconvenience. Either way, it changes his scheduling, staffing and purchasing plans. The earlier in the project you ask, the better.

Reducing scope of the project: Writing an RFQ for the maximum project, then stepping it down 20% or more after the first bids come in may not result in a 20% price reduction. Sellers can get a price in their head and start mentally spending the money, so when you reduce it, it is bound to take the wind out of their sails.

To make an analogy, imagine a co-worker told you near annual review time that the company was giving out 8% raises to nearly everybody. When your turn came you were told your raise was 5%, double the annual inflation rate. Going forward it will feel like a loss.

A contractor suffers the same effect after estimating a $50,000 project, being told he's in the final two, and then to his surprise he receives a modified RFQ to rebid that includes only $32,000 of it. Where did the $18,000 go? The same place the 3% went. It's a spoiler that steals enthusiasm for the job.

A homeowner who does this may find the two bidders present quotes for $39,000. The higher initial pricetag has a magnet effect in the mind of the estimator, making it hard to approach the re-bid as a blank slate.

They may imply you won't be happy with the curtailed size and materials, but they're simply not happy with the lost revenue. They should be telling the customer they can do a great job on the new design, but instead they may harp on how unsatisfying it will be compared to the grander vision.

If you find this happening, show him the door before his negative talk sucks all the joy out of your home improvement. His price will not come down.

Start over. Find fresh-meat contractors to bid on the new configuration. It is the only way to see the full $18,000 reduction.

Trying to reduce project cost with change notices to reduce scope or use cheaper materials may not work. A homeowner will be lucky to get 50 cents on the dollar from reductions after the contract is inked.

One reason is that the job is fully scheduled for workers and resources. If a contractor has another place to scoot his employees and equipment during the five hours no longer needed on your job, it may reduce job cost proportionally. But that is unlikely. Using change notices to reduce scope can bump up against a contractor's anticipated monthly revenue. He may have no place else to get that revenue since he turned down work weeks ago based on scheduling for your job.

The other element, materials, is just as likely to incur a cost if you change your mind. A good contractor will place orders immediately so delivery delays don't mess up the work schedule. Price breaks are based on the total sales amount from that purchase. Reducing the total materials cost could cross a price break line. If already ordered or purchased, returning might incur a restocking fee of 5% to 15% and trucking or shipping on his own dime.

It can be a shock when features or size changes that would have added cost significantly had they been change-noticed into the project are declared 'don't really make much of a difference' when change-noticed out.

You can get close to the actual reduction with good timing. During a project, conditions may arise needing extra work. Pairing a cost reduction change with additional work can result in the most benefit from the reduction. When something comes up that will add $600 in labor to the cost, you can pair a labor reduction of $500 right in the same change order to get much closer to a $500 reduction than submitting them separately. Contractors understand trade-offs, and doing this results in no cut to his bottom line.

Hiring an Architect

The architect or designer drafts the design for the GC to build. When you find one whose work matches your idea of a great house or a great addition, all you need is a builder who can make it happen.

If the architect or designer has been in business a while, he may have a handful of contractors he recommends. An experienced architect with an established, amicable relationship with this contractor is a recipe for a good outcome. You still must thoroughly check out this contractor before hiring.

An Architect will need a project target price—*not the ceiling or maximum price*! What's a good target price? Make it 60% of your maximum price. Some people advise using a not-to-exceed number, but I am not a fan of that method of controlling costs *with any vendor*. Five times out of six, the do-not-exceed is exceeded, serving as a tiny speed bump requiring mere approval to blow past. A do-not-exceed conveys that every dime below that is earmarked for him and he's a fool if he doesn't take it.

Others may dismiss this view, pontificating at length about professional responsibility. Most architects are nice people and the ones I've met seem to be ethical individuals. But people work for money. High-horse arguments to the contrary do not impress me. Dangle a carrot and good people go there. A do-not-exceed dollar figure is a nice, orange carrot.

Architects or designers are paid by one of three methods: as a percentage of total job cost, by the hour, or a flat fee. Which one is the best fit depends on the job and your comfort level. He will probably have a preferred payment method, but you could broach the subject with him if that's not your preferred method. Before doing this, only you can judge whether it's a good idea to ask, whether he may be offended, or if it risks starting things off on the wrong foot.

Percentage

An Architect working for a percentage of the total job cost gets paid in proportion to the cost of the project, and his design has a lot to do with that cost. This is the method high-end Architects use. If he has a look or style you like, then borrowing elements from his previous work is exactly what you want. Those design portions are tried-and-true, meaning any troubleshooting was done on somebody else's house and you're reaping the benefits of those earlier learning curves. If that's actually what you want, the look he specializes in, then you may not mind a percentage in exchange for a more trouble-free design.

You need to have the courage to rein him in if it becomes necessary because he has strong financial incentive to overdesign and specify costly materials that don't add significant value to you, the end user.

Hourly Rate

The second way Architects are paid, and the way most Designers are paid, is by the hour. This can be the lowest cost if there is some oversight on billable hours and he is talented. If he's less experienced the customer will be paying for his learning curve, which doesn't add value to the deliverable. Experienced Architects are the least likely to use this method, because they are quicker and better. They would make far less per job than when working for a percentage of total job cost, and in the end the customer gets exactly the same deliverable.

Flat Fee

When a customer is totally out of their element and cannot begin to guess the length of the job or what it entails, the flat fee means no pricetag surprises and you can budget for it. If you don't like doing the oversight or reining in suits who have a nicer office than yours, it's the best choice. The architect determines his flat fee by taking an average of similar jobs and then adding a bit extra for just-in-case.

A flat fee will be higher than the mid-point average for that size project and expected amount of work. On the bell curve of his work, you overpay 70% of the time. He takes a cost-averaging view of all his projects this year, covering the higher-labor-than-expected jobs with money from the as-expected ones. If this job is more work than the amount of hours he calculated, he should be able to absorb it with the just-in-case money from other projects and still make a good living. If he's just so-o-o good that almost everything he touches goes smooth as pie and he makes out like a bandit, well, that alone is worth a couple extra grand!

I discourage hiring contractors at a flat fee, but architects are a different matter. The bell curve of variation in their workload is far less. A site situation entailing changes to the design might take an additional five hours of work for the architect but forty man-hours and $900 in materials for the builder, so an architect will stick to his flat fee under circumstances that a builder would not.

The biggest plus to the flat-fee is there is no incentive to over-engineer; he can be a good ally in keeping construction costs down and work progressing since it's no more money in his pocket when construction costs go up.

A flat fee guy is the most motivated to keep start-to-finish time to a minimum and is most likely to call out specs at the best value per dollar. It's almost certain he has a good amount of experience, although it's possible your project may be conducted by his employees with his oversight. That may happen with any payment method so if the individual who serves as the primary is important to you, have that discussion and write those verbals into the contract.

Speaking about human nature again, the flat fee guy is the most likely to lose steam, begrudging effort and time towards the end of job. Unlike the other two who will financially gain from hours spent solving issues or making small improvements in the last weeks before completion, this fellow could lose all his profit margin throwing extra time at it. This risk is not most of the time, just more likely than fellows paid one of the other ways. This means if issues arise towards the end you may need to be more of a squeaky wheel to get his attention.

* * *

The choices are: flat fee, with worst-case risk money worked in but most likely to be a partner in keeping contractor and materials costs in line; a proportional percentage, where the best in the business usually hang their hat, or hourly, paid for every hour devoted to your project.

Add On or Move?

It's common to start with a modest addition, say a bump-out of the living room and master bedroom above, and then add a new bathroom, and then add an electrical system upgrade, or any of a dozen other wonderful improvements. Planning an addition can be like the proverbial frog in the pot on the stove: the water heats up so gradually that at no moment does it decide it's too hot and jump out, and thus gets cooked.

Don't get financially cooked by incrementing your addition past 25% of your current house value. In every community, a house will increase in value up to a maximum of 10% more than the neighborhood average. Trying to get more at sale time means your house will languish unsold. If getting 50% of the value of the improvements means the house price would be 10% above the neighborhood average, do it for your enjoyment alone.

If the pricetag on your project goes higher than 20% of current value, seriously think about buying a different house with everything you want in the original design, not as a tacked-on afterthought. That house might be only a block or two away.

Online, it isn't hard to find bald-faced total lies about the return on investment for various home improvements. A few: installing a steel front door will recoup 70% of your investment and a new deck will pay back 60% at home sale time. If everyone in the neighborhood has a deck, your house will be even-steven. It can lose value for having no deck or having a junky deck, but it can't soar another $5,000 over the neighbors for just having a new deck. Appraisals don't work like that.

A door is a door to an appraiser, even if it's a spectacular door; there's no allowance on his standard form for anything more than 'in great condition.' You can get the same rating with a coat of paint, new kickplate and new door knob. One household project entailed replacing an awful fireplace made from adhesive-backed plastic 'bricks' and a 2 x 8 board for a mantle with a new stone front going to the ceiling, a 5" thick solid oak mantle and a one-piece granite hearth. When it was done, as far as appraised value was concerned it was still a fireplace.

This fireplace, however, wasn't valueless. Home improvements can be split into two types: those that increase the specs and those that reduce days-to-close. Wow-factor improvements such as a stone fireplace, a spectacular front door, an elegant patio, big windows, California closets, or a built-in china cabinet appeal instantly to buyers. An outstanding feature or two makes everyone overlook or minimize other drawbacks.

Upgrades such as adding a bathroom, bedroom or garage, finishing a basement, or installing central air conditioning become a new appraiser line item so will increase the value up to the high end of the neighborhood and then flat line.

Adding a basement bedroom or one bath may attract a different set of buyers but within that set you're competing with homes having that fourth bedroom or second bath as part of the original design. Think about the competition. It might be easier to compete as a 3-bedroom with a gorgeous kitchen remodel than as a 4-bedroom with one of them being damp, having a low ceiling and a cold floor.

Don't assume the improvements are like cash in the bank unless the shabbiness is sinking the value. If you're determined to do something just before sale, focus on improvements that reduce days-to-close. A local real estate agent can rattle these off for you in your price range. Don't guess at them because they're often not what you expect. Regional differences are significant.

When you sell a house, you sell at the neighborhood average. To get buyers to pick your house out of the seven they looked at you need one to three features that are hard to find at that price range. For days-to-close improvements, the way you get your money back is in freedom to move on your schedule, without overlapping mortgages, bridge loans and tax bills for a few months or worse, living in a motel in another city until the house sells. Expenses related to long days-to-close can amount to thousands if not tens of thousands of dollars.

If you're planning a major renovation or addition, when costs have crept up to an unintended level, pause. Put a pin in it and bring 'buying a new house' back on the table. If you're contemplating an addition on a 1200 sq. ft. house to make it an 1800 sq. ft. house, just buying the 1800 sq. ft. one will leave you in better financial standing two years from now than adding on.

There are exceptions. Some of those exceptions could be if the value of your land is high, if the neighboring homes are mostly 1900 sq. ft., if you just can't sell this house, or if the prevailing interest rates are several percentage points higher than the main mortgage. If the main mortgage is fixed at 4% but the best mortgage rate today is 11%, then an additional mortgage for only the improvements is sound thinking.

The costs involved with moving are more predictable (not perfectly predictable) than the costs of an addition. People who know construction will quip "Calculate the cost of the addition, then double it." No one says that about house-buying.

The worst reason to become committed to an addition is a dread of the hassle of moving. The hassle of living through a drafty, noisy, disruptive, messy construction project is usually underestimated and can be worse. If you don't want to move because you like your neighborhood and your commute, consider becoming someone who moves only a few blocks into a house they like better. There are a lot of people who have done that; I know, I am one.

The question should be, "what's the best way to have the desired features in my abode and be the most financially robust in two years?" The answer to that may not involve living with a tarp for a wall for three months.

Proceeding with a Renovation or Addition

Cost overruns are the name of the game. Do not believe for an instant there is anything like a flat fee from contractors. All contractor work is time and materials. Of course, all bidders will give you a price, a dollar figure, for the project. Your contract must contain a total price. If that ends up to be the price you pay and no more, and you get exactly what you expected, it's because he's a very experienced estimator AND no big complications were encountered.

If the job encounters extra work not included in the initial quote, when it runs beyond the risk factor money[21] he added, which is his little slush fund so he doesn't nickel and dime you, then he will ask you to fund the cost overruns or he'll stop working.

The contractor's other alternative is to make ends meet by not paying suppliers, completing the job for the agreed-upon price but leaving you with suppliers who demand you pay them. Either way it's more money, and the lesser of the evils is your contractor dinging you for extra. You must get those lien releases!

Because you shop by price and may pick one guy over another for a $1,000 difference in a $50,000 project, a contractor has strong incentive to make no attempt to foresee complications. These will all be handled by cost overrun.

When your contractor discovers the ground requires deeper pilings or rot in the adjacent walls requires more extensive tear-down and rebuild, keep in mind that it wasn't caused by this contractor. Every single one of the bidders would have encountered the same thing and forwarded the cost to you. That's why you must allow some holes to be poked in walls, ground tests, and whatever it takes so if these things exist, ALL your bidders have, as part of their competitive bids, the deeper pilings, rewiring the entire house, fixing all the rot, replacing the sewer pipe, and all that other stuff which usually ends up in cost overrun.

Having these in the initial bid not only means you can compare quotes apples-to-apples, but it lowers the cost. When defects are found smack in the middle of their workday, work comes to a full stop. Different equipment must be rented and brought to the site. Someone has to go to the store. Had

---

[21] For a full list of risk factors, see *Addendum B, Ten Risk Factors* that increase the price.

this been part of the original plan, no work stoppage, no renting of two sets of equipment, no lost days of work.

Take, for instance, discovering more extensive wood replacement is needed. He already bought all the lumber for this job. Now he has to go buy more, and make a second truck trip. Had he bought it all in one trip he would not only have gotten a better quantity price break but saved gas and the labor of the fellow who takes three hours to load it into the truck and drive it over.

I know you hate, hate, hate to fish for bad news right at the quote stage, but force yourself to fish. Get as many of the 'complications' as you can included in the initial quote, as painful as that feels. If you wanted a $45,000 addition, with the additional foundation work and electrical work it might come to $51,000. Ouch. But if you sign a contract for $45,000, when those things crop up and the BI will never sign off on the job unless they are fixed, the work stoppage, change-of-plans way of doing it will cost you $54,000. Ouch-ouch.

To get the job, contractors tend to quote just the basic work. They know, however, that the older the home, the more likely it is to have mystery wiring, circuitous plumbing, foundation problems, and cheating on code. Very old homes are almost certainly totally out of code, everything from distance between studs to unsafe you-name-it: floors, ceilings, and everything in between.

Yet the initial quotes for remodeling and home additions may account for none of this unless you demand efforts to ferret out existing defects so the cost is included in the initial competitive bids. Have a guy come out with fiber optic scopes and camera equipment to peek through walls, down drain pipes, under floors, all over to find those hidden defects. Have the ground tested to determine true support needs.

Although to the first timer it seems like anything can come up, the same problems become routine in their regularity to those doing renovations or additions.

Who could possibly know ahead of time that a Massachusetts house built in 1923 would have rot in the beams? How could it be foreseen that a beach house converted to year-around living ten years ago has bad foundations? Who could guess that in a 1960's subdivision where they built twenty-five houses in one summer, the house wiring is not to code? Who? Every GC with a year in the business, that's who. These are going to be the case 90% of the time. It's a pleasant surprise if they are not there.

Ask your contractor to speculate on what kinds of things could be problems in your project, and then ask what can be done to either confirm those or totally rule them out.

Great Home Features for Pennies

When building an addition, a major home remodel or even a new home, there are many wonderful touches that builders don't instinctively think to add. Architects may believe some fine points are beneath them to include in the detailed plans, and the GC isn't going to ponder your day-to-day living comfort while in the thoes of handling staffing, materials coordination, BI visits, subcontractor hiring, and all that. If the quality improvement would be a daily joy to you but cost him $6 more, it will never magically happen.

The only person who can make these things happen is you. If you don't, builder's cheap will be installed.

Builder's cheap is my shorthand for the absolutely worst quality, lucky-to-last-five-years crap that builders install when left to their own devices. Things like $1.99 porch lights, $5 can lights, $1 wall outlets, $3 doorknobs, $6.99 bathroom faucets –these things are so cheap that the big box hardware stores won't even carry them, they're so awful. Builders buy them in boxes of 20 and install them everywhere.

Pick your own faucets, lights, doorknobs, switches and so on; don't surrender no matter how much your builder protests. It's likely to boil down to him not giving you credit, meaning no knocking something off the price which initially included these things. He's too ashamed to admit what he was going to install tallies up to $93.87. If the faucet you pick runs $90, don't expect him to deduct $90 from his price, since he was going install a $7 one.

Do it anyway because the labor portion is more than twice the price of the item, and replacing builder's cheap in piecemeal fashion over the next eight years will cost you thousands. It's the same labor to install a $7 faucet as a $90 one, but the latter will look great and be the room's focal point for fifteen years, while the former will leak and look bad in three years. Porch lights? Closet doors? Those will be embarrassing, meaning you won't bear them for even three years. This chapter contains the full list of items to select for yourself whenever they fall within the scope of your project.

You cannot be 'too busy' to do this. Unless your time is normally worth $200 an hour, getting these done right the first time will save you this much in rework over the next four years.

Pick These Yourself

Make a champion attempt to select most or all of these for yourself:

>   overhead light fixtures
>   fireplace surround
>   mantelpiece
>   front door
>   screen doors
>   inside doors
>   closet doors
>   doorknobs & door locks
>   doorbells
>   light switches
>   bathroom lights
>   bathroom exhaust fan
>   can lights
>   all external lights
>   bathroom mirror
>   sinks
>   faucets
>   trim

These can improve enjoyment and reduce home expenses for years. The price difference between cheap and high quality closet doors, for instance, is $40 a set, so upgrading six sets in the whole house could be only $240 more.

Upgrading a faucet at first installation could cost $70 more; having a plumber replace it five years later, $290.

Fireplace surrounds: upgrading by $500 can make an astounding difference that will change the impression your guests (and future homebuyers) have of the whole house. There is so much bang for the buck in the appearance of a fireplace that it is sad when a builder or homeowner chooses to save $200 or $500 right here.

If the contractor insists he buys in bulk and won't custom purchase these things, one way to break the impasse is to offer to buy them yourself and bring them to the site in exchange for a free upgrade on something he does offer. Getting a free $500 upgrade on say, a skylight, can take the pinch out of buying faucets, light switches, doorknobs and screen doors yourself.

This chapter also contains great features that the boots-on-ground workers can make happen with measuring and cutting choices, meaning these are free. You must request and demand these low cost, even no-cost quality improvements in a bold print, three-times-repeated manner. I'd like to tell you that you could read this chapter, jot down a list of all the good features you like, hand it over to your builder or include it in the RFQ, and you will live happily ever after.

It has never happened to me yet, so it would be wonderful if you are the first. The way to get these is to write them in the contract, then remind and pester and put your foot down and never waver.

Builders often focus their efforts on winning your job as the low bidder. After he has it, he will press you to abandon the user-friendly upgrades in your RFQ. It will take fortitude and strength to stave off this blatant profiteering and stick to your contract and your RFQ.

Expect him to descend to being dismissive and even insulting to get his way. Stick to your guns and your contract when you call for better materials and he wants to use cheaper ones.

You should respect his competence and digest his advice when your design ideas are, shall we say, a bit too unique, or the cost of doing it that way will be unnecessarily huge. On the next pages are many great features that, in my opinion, someday building codes may adopt as standard, but until then you will have to add them into your contract yourself. As mentioned in the *Watching the Work* chapter, these are things that will need mentioning to the crew on the first day of work and at key times during the project, just before they do the wrong thing which then has to be torn out.

It will not hurt anything for workers to hear something twice. Or even four times. It hurts a lot to do it wrong and then have a spitting contest over who pays for the labor to tear it out and do it over.

Outlets

The lazy way is to put all outlets the same height from the floor all over. In some places like the laundry room, home office & bathrooms, four feet off the ground is more practical.

Where the electrician plops an outlet is literally somewhere here and there about six to twelve feet apart. He puts them where it is convenient for himself at that minute. Do you have any extension cords in your home? That's because the outlets aren't where you need them.

It should cost nothing extra to request a four foot high outlet where your desk will be or move one over a few inches to ensure it will hide behind the TV stand. Ponder if a location would be better served with a floor outlet, baseboard outlet, inside-a-cabinet outlet, waist or shoulder-high outlet in built-in bookcases, or ceiling outlet. Outlets don't need to be shin-high in a wall, they just usually are.

In large rooms, i.e., great rooms, it is always extremely useful to have at least one floor outlet where furniture will be clustered away from the walls.

Tell the electrician where you want outlet locations and heights. The electrician may balk because he 'has to' put an outlet right here. Fine, but he can put another where you want it, it doesn't have to be either-or. You are the customer and this is where the rubber meets the road if you're an electric-using person. Speak directly to the electrician; telling the GC where you want outlets, even putting marks on the walls has about a 40% chance of compliance.

A written note or sketch showing locations is always good. Being there face-to-face with the electrician is preferable; if you can't, then giving something on paper to your GC has a good chance of being passed on to the electrician. Always include your phone number on notes to contractors or subs.

Living room outlets: They can be placed in the baseboard. Baseboard is usually hardwood so it's a bit more time-consuming to get it right. It may be too expensive to do in every room, but if you have an elegant room, it's classy to recess a wooden plate in the wooden trim so the walls are pristine. Plates can be painted or wall-papered to match. Consider floor outlets too. I'm flabbergasted how often an outlet is located where it will be unsightly forever. If you can see this room has a location like this, don't depend upon the electrician's decorating sense telling him not to put an outlet there; show him exactly where to put outlets to avoid having one in an area you want unmarred by unslightly outlets.

Bathroom outlets: most have one outlet near the sink. In today's larger bathrooms, consider putting another across the room. The second outlet can be used for many things: phone charger, electric bathroom heater, radio, TV, coffeemaker, or all of them. Install the outlet a few inches above a wall-mounted shelf or a tabletop in that area. Tight space? Put the shelf 6'6" high and the outlet 3" above it, where it will serve nicely for a wall-mounted TV, radio & charger.

Closet outlets: An outlet in a closet is wonderful for keeping unslightly hand vac and cordless drill chargers out of sight. It keeps chargers off the kitchen counter. Closet outlets should be at chest or stomach level and within arm's reach of the door.

Outdoor outlets: If you do outdoor decorations at Christmas, request outlets located so the cords do not have to cross sidewalks or go over doors. Pick the locations because simply asking for 'two more' outlets will not work; the electrician puts things where it's convenient for him with no thought for its use. In the back yard, additional deck or patio outlets permit coffeemakers, blenders, mini-refrigerators, table lamps and the like placed on tables during parties.

Non-allowed outlet locations: Get outlets where you want them and as many as you want. Sometimes an electrician or contractor will say he can't put an outlet that close to a window or near a sink due to code. He's fibbing; he only needs to put a GFCI outlet there, which costs $10 more than a plain outlet and needs a little more care with the wiring. If you run into a fibber, just say "Could you please install a GFCI?" It's worth engaging the argument because you're the one who will live with a trip-hazard extension cord running in front of the patio door for the rest of your life if you allow him to skimp on $10 measly dollars like that.

Garage outlets: These usually need to be GFCI. Most new construction puts one outlet in the garage. Most people put a workshop with tools, extra refrigerator or freezer, or even a hybrid car in their garages. All reasons why one outlet is insufficient. The cheaty way to get an extra outlet is to buy one of those light switch/outlet combos that fit into the same space the light switch does. Your handyman can install it, or you can do it yourself by following the instructions in the package.

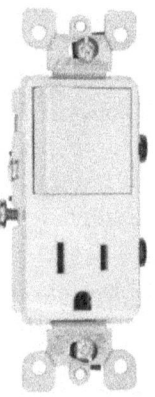

Adding an outlet where currently there is a light switch

Phone, Cable and Internet: There are modular outlets that can be wired to provide two to six ports in one plate. It's better to install the modular plate during construction even if all three won't be hooked up right now. Place a modular outlet with all three ports in the living room, family room, kitchen and master bedroom even if you're certain right now this room will never need it. If you have a phone jack or cable outlet in a location right now, instead of poking another hole in the wall to add another plate 4" away, which is what the average phone or cable guy does, purchase a modular plate for the poked hole already there and run all the wires to it. You'll need to purchase the plate and jacks yourself and have them on hand for the service installer. I've never seen a service installer suggest these, but how wonderfully competent if yours does.

Modular plate; coax (cable), Ethernet, phone, HDMI, audio or USB can run from room to room.

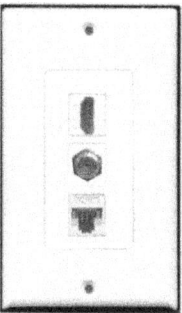

Unit preassembled with HDMI, cable and ethernet, $16

Light switches

There are light switches that have an LED in them and glow a bit in the dark; these should be mandatory in spaces like front hallways, bedrooms, kitchens, basements, garages, bathrooms without windows. . . come to think of it there are few exceptions. Any location where someone might fish for the switch when the lights are off is a good location. These cost an extra $2-$3 a pop. The lighted switch is a never ending joy. Get them.

Light switch locations are often given no thought. There is no rule that a light switch can't be installed next to the bed in the children's room. What's the downside of having three light switches in the children's room? None. The light switch is usually put on the inside wall near the door; for a dining room, why not on the wall outside the room too? Why should anyone walking in have to fish in the dark for the switch when it could be in the lighted area before they go in as well as inside? Doesn't it make a lot of sense to have the basement stairs light switch outside the door to turn on the light before you open the door?

Hallways: people often buy battery-operated motion sensor lights they affix to the walls at knee or ankle level. Consider having small motion-activated lights installed in the baseboard. Alternately, your electrician could replace the light switch with one having a built-in motion detector so the light comes on automatically when it's dark and someone approaches. Dimmer switches are often installed in locations where they are kept on high all the time, but the hallway is one location where a dimmer switch in conjunction with a motion detector will actually be valued.

Forced Air Register Vents

It's common to see a register several inches from a wall but just short of sufficient gap for the average 11" wide vacuum cleaner head. Moved just an inch or two over would make vacuuming easier. The hole for the air underneath the register is between 8" to 10" wide but registers are 18" across, so moving the register over 2" is not only possible, it does not affect duct location.[22]

End tables, curio cabinets and the like are often about 24" wide. In corners and along flat walls it would look better if the furniture fit in the space beside the register rather than having one leg right in front of the register. So if the distance is over 21" from the corner, make it a 24 1/4" space, and if it's over 33" make it a 36 1/4". In most rooms the centering of registers to wall features is irrelevant, but furniture sitting tipped or not against the wall because of the register will be a constant annoyance.

A third common vent error is placing it right beneath a window. Seventy years ago when 'drafty' and 'windows' went together like salt-and-pepper, heat vents right below the window might have mitigated it. Window technology has advanced beyond that. Forced air vents below a window cause the drapes to balloon out, impeding air flow into the room. A vent under a window limits window treatment choices and is the logical place for a chair, table or plant stand. Locate vents to the side of the window, clearing it by at least twelve inches to allow drapes to reach elegantly to the floor.

You must say something about vent locations early. Once they cut and nail in the trim, it is a huge time and wood waste to move them, since the trim will have to be discarded and the wallboard might be damaged during removal.

Doorknobs

Most builders know better than to cheap out on the front door doorknob, but not all. To guests coming to your house, their first impression is the curb appeal and then the front door. The front door handle provides a huge bang for the buck.

---

[22] While radiant heat, meaning heat coils under the floor, have their strong advocates, it doesn't free the house from ductwork because the AC still requires ducts. All new construction and major renovations should include central air if it isn't already there, or risk a serious hit to house value. With today's more air-tight and insulated homes, forced air systems reduce odors and maintain a healthier environment. While AC-only units 'could' go in the attic and be distributed via ceiling vents (cold air sinks, heat rises), this tends to lead to premature failure of the expensive systems.

Online vendors and hardware stores have hundreds of beautiful matching sets of deadbolts and doorknobs. If you can't find both handles and deadbolt that you want in a matching set, a locksmith can change the locks inside your chosen hardware (from the same manufacturer) so the same key works on front, back and garage side doors.

Deadbolts: On exterior doors, the vast majority of builders do not immediately install a deadbolt in new exterior doors, leaving that for the homeowner to do later—which is silly, because every home-owner wants one. Installing a deadbolt while they have the tools right in their hands takes an extra fifteen minutes.

The front and back doorknobs can be different colors and styles than the inside doorknobs because they're not visible simultaneously with the interior doors. Entry doorknobs should match the outside look of the house and the level of security you want. Make no effort to match the style or manufacturer of the handles on interior doors.

A word about locksmiths: this trade is becoming a hotbed for thieves. Statistics are bandied about that up to 90% of the website listings may just be fronts for unscrupulous people. Beyond the obvious, which is that they have a key to your place, some arrive and then jack up the price by hundreds of dollars. It's better to scope out a locksmith with a retail storefront before you ever need one, go in and say hi, and if it feels on the up and up, put his phone number in your speed-dial and give his number to all your friends too. We have to help the good guys in this field.

Inside doors: The lever-style handles are pure luxury for interior doors as well as exterior. The first time you open the door with your elbow because your hands are full it will be worth the few extra bucks. Helpless resignation to all the negatives of round knobs is the only excuse for buying them. Despite the huge variety of round knob designs for sale, in daily life we hardly ever look at them. Quick—in the last three houses you visited, were the knobs gold, bronze or silver? Realistically, they are not memorable. The only extra dime worth spending on them is for ease of use. Get levers. Functionality is what adds to quality of life.

Attic

The location of the attic access opening is usually in a hallway where it remains an unsightly air leak forever, with its little red pull-down handle dangling. If the house has an attached garage, have the pull-down attic stairs installed in the garage; it all goes to the same attic. Don't allow the builder to just close it with a board, make sure he installs the pull-down stairs and adjusts them to the right height too. Putting it in the garage eliminates heat loss in the winter, and may even be more convenient and accessible for things like seasonal item storage.

For a 2-story house or one without an attached garage, stairwells are a good place. Ideally, when the steps are lowered they will rest not on the top landing but one or two steps down from the top, providing more head clearance for the user. Also, and this is important, the opening should not be centered on the stairs but skewed to the wall without the handrail. This allows someone to squeeze by while the steps are down. It also means when carrying or handing things up, a person can brace their back to the wall for support. There's no chance of falling off into a gap.

Garage

When building a new garage, ask for a bid on drywalling. The cost to install the good $12 a sheet moisture-resistant drywall in a 2-car garage could be around $600 to do it immediately after the framing. Materials: $240 + 2 guys, 3 hours @ $50 hour. It's cheap because big sheets are nailed up with very little cutting, no impediment to material moving, and super-close access to the truck. Doing it later will never cost that little. Ceiling and painting are extra.

Please reconsider if you're thinking of saving money by not drywalling; arriving and leaving the house while looking at that raw framing is bad feng shui. The room where your car sleeps will be so much brighter, not to mention won't become a bug hotel, if you drywall.

Another thought: for a little extra money today, you could double the usable space in the future. If there will be a drywall ceiling, have them also install a sub-floor and a pull-down or permanent stairs to the attic area.

Request a vapor barrier in the ceiling to block car exhaust (code-required) and have an electrical outlet or two plus an overhead light installed upstairs.

For a bit more money, ask for a window to be installed on each of the flat sides, if it has a basic upside-down-V roof. For a bit more, have them make a dormer facing the back yard with a few more windows. Any additional work can be done later as needed.

Every state has its own building code rules, so check to see if there are additional rules regarding rooms above garages or rooms in an attic that apply to your situation.

Plumbing

"Plumbing" in a home is actually two separate systems: slender hot-and-cold clean-water pipes that push the water under pressure, and larger drain pipes that work by gravity (except when an electric pump is involved) so need to maintain a certain slope to exit the building. They were not mistakenly installed crooked, which is a common impression.

A city or township usually pressurizes the water supply by using a tall tower. A smaller pump works 24-7 pushing water up into the tower. When a customer turns on a faucet or flushes, gravity is creating the pressure. If everyone takes a shower at once, gravity provides the pressure so everyone gets a good rinse. Then the pump works to refill the tower. People with wells, communities without bubble towers, and tall buildings use an electric pump (usually underground) to provide pressure. You'll know which one you have when the power goes out; bubble-tank people can still take a shower, pump people cannot.

The drain portion of the plumbing system will work fine without power for those using only gravity. There is pump-assist drainage also, due to hills and valleys within a community that make it almost impossible for the sewer system to be lower than every single house served by that waste processing station. The pump-assist people can fill their drain pipes if the power is out for a long time; for most short power outages there's no risk.

Sewer backup, a term I'm sure you have heard, is when the drain pipes have a blockage or are just plain full. If your sewer lines are connected to the main line and not to a leach bed in the back yard, anyone uphill from you that is taking a shower may not know the line has filled to two feet below his drain, which is two feet higher than your toilet lid. Leach bed folks can get a sewer backup too, during heavy rains.

Most of the time you haven't done anything wrong to get sewer backup. The street grates that catch rainwater go to exactly the same sewer pipes as the toilet flushes and sink drains. The exception is when tree roots and other debris clog your drain pipe between your house and the main line; this requires a Roto-Rooter service to clear.

Inefficient plumbing routing is almost standard in older homes because some or all of the plumbing was added later; often the house was not designed with plumbing in mind.

When renovating a pre-1960s home, and even some after that, it's fair to assume this project will require some plumbing work right back to incoming cold water pipe or outgoing drain. Reasons even a post-1960 home could require plumbing repair include: rusting out, mineral deposits narrowing the inside of the pipe, inadequate past repairs, stress or bends that will eventually break at that point, not-to-code materials, or no shutoff valves installed. Or it could just be Roto-Rooter time.

Preventing Frozen Pipes

This is so simple that it feels presumptuous to tell plumbers and GC how to do it, but it is so often not done that you may just need to insist upon it to get it. They will make wrong decisions just to save $12 in elbows.

Plumbing lines running beneath the floor or between floors should not run along outside walls. Elbow them in about three feet and then run to their destination. Even if you think you live in a generally warmer area, as the folks in Atlanta did until the winter of 2013, 'generally' does not mean 'every single minute' and it only takes five hours of cold to burst a pipe.

The primary cause of freezing pipes is a pipe running along an outside wall. Heating the house will have little effect on a pipe within 12" of the outside wall, especially if that pipe is sandwiched between floorboards and a basement ceiling, which act as insulation *against* the warmth of the room. That sliver of space between the floors is closer to the temperature of the nearby outer concrete wall.

The best way to prevent pipes from freezing is running the water. Most vulnerable are sprinkler-system pipes and less-often used pipes, like the line to the washer-dryer or outside faucet. Sprinkler systems should be built with a bleed into a drain on the far end so the same water doesn't sit in there for decades. When you hear of a business that has suffered water damage from burst pipes, it's nearly always the sprinkler system overhead.

The old farmer trick is to keep the kitchen faucet at a trickle all night. Running the water pulls the underground water, whose lines are buried deeply enough where it's always 50°, into your pipes exposed to air temperatures colder than 30°. Bringing this warmer water into your lines not only breaks up any forming ice crystals, but moving water has a lower freezing point than still water. Flushing a toilet every three to five hours plus running the farthest-away faucet for a few minutes every few hours will keep you frozen-pipe free.

Basement

Having an egress window in the portion of the basement where people might be watching TV, sleeping or playing may be required in your state. The size of an egress window is not the minimum size needed for people to get out; it's sized so a firefighter with an oxygen tank on his back can get in. When putting in foundations, installing an egress window costs a pinch over the price of the usual basement window. Once you have it you'll love it; it lets in natural light, making the space downstairs more livable. Outside, a thin clear plastic rain shroud costing a few bucks prevents it from becoming an aquarium.

Roughing in plumbing pipes for a future bathroom downstairs is less important than it was several years ago due to technology improvements in pump-assisted up-flushing toilets. There are also units called sewage-ejector systems that have a flat tank beneath the floor so the bathroom floor will be about 6" above the cement. With this unit you can use any retail toilet. Up-flushing toilets and ejector systems make a bit of motor noise and require electric power, but they do the trick. Additionally, a sink and shower can drain to it, so one pipe connecting it to the home's sewer system suffices for an entire downstairs bathroom wherever you want it.

These systems can range from $600 to $1,200 before installation, and they eliminate the need to jackhammer the basement floor. Concerns regarding drain angle or distance from the street for the lower-level toilet, which used to limit where the bathroom could be located, disappear when using these units. Anyone can have a full bath in the basement without cementwork. Avoid first-timer installations; find installers who have put in your particular up-flushing system a few times before because there are a couple of ways to do it wrong.

Tile

Picking a nice tile is 70% of your happiness with it; the other 30% is how they place it. Leaving your tile-layer to his own devices means most of the time you'll wish he had done it another way. So look at how the room is used, what spot you'll look at the most, and command them to place full tiles in that spot.

On floors, backsplashes and bathroom walls, there are two ways to go: one, lay a full tile 'in the middle' and work towards the walls. This means the end cut tiles are symmetrical. If you have 12" tiles (with ¼" grout between) and the room is eleven feet wide, approximately 1.5" will be cut off each of the end tiles.

The other way is to observe where the eye lands upon entering the room or the most-revealed spot while in the room, and tuck a full tile into that corner or edge. Then the tile layer works to the other side, cutting only the tile on the far end as needed to fit.

The second way is absolutely the ONLY way to go if one edge is behind the toilet, under a kickplate or overhang, in a closet, under a garbage can or behind a door that's usually open.

Tile layers get very devoted to the first way, beyond all reason. What makes it worse is 80% of the time they are very loosey-goosey when measuring, so while they tried for the ideal of symmetry, they end up with one side having a 4" wide tile and the other a 2" tile. Yuk. Symmetry only works if they are accurate to within less than ¼". You will have to request accuracy and better yet, a second measuring, if you go the symmetry route.

Case Study: A living-room project to tile the front door foyer area plus the adjacent coat closet to the left. The tile store provided an experienced tile-layer. Two sides of the new foyer abutted the living room flooring. The tile layer wanted to center the tile along the width, which ended inside the closet on the left. This meant there would be a partial tile row in the living room. It was a battle to make him to lay only full tiles in the living room and put a partial only inside the closet. He protested only centered tiles look good. No common sense, doing his job on auto-pilot. Don't expect more from your own tile guy.

Case Study: Spa shower with 12 x 24 tiles in an aligned pattern (not staggered). The shower was on the far wall of an L-shaped bathroom where only one corner of the room was visible when walking in. The tiles extended past the shower to the opposite corner located in a toilet nook. The very experienced tile guy was a hard sell, but I got him to use full tiles on both walls in the shower corner and cut partials only at the far ends. Once done, it's obvious the stunning impression would be diminished with a row of halfsie tiles right at the visual focal point of the room.

Of the three tiled rooms in my former house, two of them had only one corner that was visible upon walk-in or use; yes, odd-size cut tiles in two planes in that corner in both rooms. Not thinking about focal point is the default when tile-laying, unless you do it for them.

Tip: if this is a shower with a drain and you have large tiles, it is really nifty if the drain is smack in the center of a tile, and then the floor slopes to it. Or two tiles, each cut exactly in half. If this doesn't work with your tile choice, have a discussion on how to make the tiles around the drain look good; broken, tipped and badly-located tiles around a drain are the norm. In tile showers, excellent-looking around the drain is the number one priority.

Bathrooms

In most new construction, both architect and builder have bigger fish to fry, so they install bathrooms that are blah. They install sinks, toilets, tubs and showers that someone spent twenty whole minutes picking from a catalog. If yours is a half million dollar renovation or addition, to look luxurious they will pick gaudy and busy-looking woodwork, add a few pillars, mix three types of expensive tile with swirls, and top it off with some huge, ridiculous tub that takes 40 minutes to fill so gets used three times a year on birthdays and special occasions.

What you need in a bathroom near the bedrooms is

1) storage space. Some of it should be within reach of the toilet.

2) a big mirror and good lighting.

3) a sink with space, hidden or in the open, for sprays, soaps, lotions, medicine, brushes, toothbrushes and razors.

4) a shower, optionally a tub, that protects users from viewers outside the house, and possibly from viewers in the room.

5) a robust air exhaust unit.

Two other possibly personal preference requirements: the bathroom must be sized so it can be warmed up quickly via heater or blow dryer, and never a window exposed to shower spray. That indicates poor design. A window exposed to daily showers rots. It can't be opened while showering and still protect modesty, and when closed it must be covered with foggy glass or a curtain so it provides no view. Can't open + can't look out = use glass block instead. Back on track, it's up to you to coordinate your bathroom look, not simply accept builder's cheap. The best way to do it is to look at lots of bathrooms until you find one you like, sleuth out where those parts were purchased, and . . you get the idea.

To start, you can use Google. Look for the word 'Images' on the top right and click it, then type in 'bathroom designs.' There are even more on iStock.com, the source of the bathroom views in this book. Screen shot or copy all the tile colors, sinks and consoles, towel racks and so on, that you like. Now all you have to do is shop for similar colors and looks, or show the photos to your builder and have him match it. Better yet, make the annotated photos an Addendum to your contract.

It is backwards to pick out components and tile at the big box store and hope they look OK. Start with a look that you find relaxing and useful, and choose materials that take you there.

Bathrooms are a good candidate for heated floors. Another option worth considering is a wall heater. There are many nice designs for under $200 that are sunk into the wall or mount on the wall and protrude slightly. Once you experience warming the bathroom while showering, you will wonder why you waited so long to install it.

There are wall-mounted electric fireplaces which may not only heat a very large bathroom but will add to the extravagance.

Ceiling exhaust fan. Upgrading it at initial installation by even $30 can make a huge difference in noise, lighting and air flow – things you will notice every time you use the bathroom.

When a bathroom faces other houses or the street, consider a column of glass block instead of a window, fitting between two studs, two blocks wide, and extending from floor to ceiling. It provides the maximum light without fussing whether someone can see in or not. Today, most homes have AC and exhaust fans so few people open the bathroom window anymore. If a venting window is important to you, install a transom window above the glass block. It will be better at venting heat and humidity than a window that cracks open at waist level. It can be wide open while you're still undressed.

Pick your own faucets. An additional $50 or $75 over rock bottom gets you an amazing faucet.

Upgrading the sink vanity is an excellent value. An extra $200 here can take you from a ho-hum console to one that impresses. What you can find online for under $500 is vast. Don't accept builder's cheap.

The height of the sink vanity is a matter of personal preference. The standard (low) height is still advisable for bathrooms used by children, but consider raising it a few inches in the master bath or 1-bedroom apartments. The standard height was chosen decades ago, when the average woman was 5 feet tall; now the average woman is 5'5". Face-washing and other tasks are more convenient when it doesn't require bending over much. Standard height cabinets can be made taller by using a taller kickboard. Wall-hung cabinets can be any height you like.

Even small bathrooms - this one is only the width of the tub - can be stunning for a few hundred dollars more in fixtures and the right tile choice. High windows let in light and require no blinds or drapes. Mirror-front cabinets make the room look larger.

Pocket doors on master baths solve the door swing problem.
Photo courtesy of Johnson Hardware, johnsonhardware.com

A double column of floor-to-ceiling glass block between two studs provides both natural light and privacy. Toilet and sinks are wall-mounted for easy floor cleaning. The wall bump-out by the toilet hides the tank.

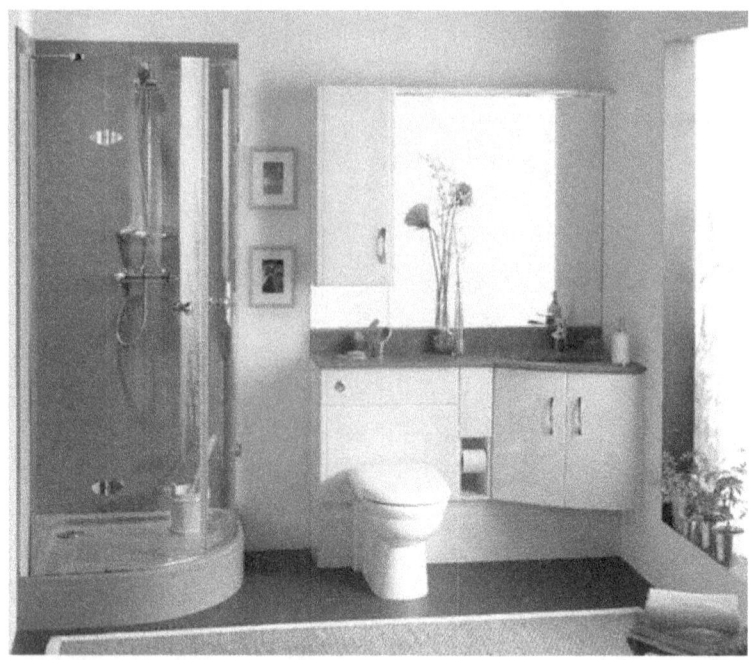

Anyone can make a large bathroom beautiful; sleek design in long, skinny bathrooms is harder. Creative cabinetry provides storage and functionality. *Photo courtesy of a-v-designs.com*

A tile floor with a slight slope to the drain permits a no-door shower in a dinky bathroom. A more attractive sink and lowering the mirror a few inches would make it even better.

Powder rooms are really simply for the toilet. Consider installing the smallest sink you can find and making the floor the focal point. Mount a shelf above the door to hold spare toilet paper and towels. *Photo courtesy of femtalks.com*

Steps and Stairways

A best practice for steps and stairways is to not have any. If doors on the addition, patio or entrance can be reached with slopes that qualify as wheelchair accessible, do it. Once you do you'll be surprised how handy it is. Building code governs the maximum slope so there could be variation from state to state. Basically, there are two slopes: a normal shallow slope and the steeper wheelchair ramp slope. The latter requires a handrail. A normal slope is 10" per ten feet in Massachusetts. A slope of 6" per ten feet is legal everywhere, is hardly perceptible yet can eliminate three steps on a thirty-five foot walkway.

Step elimination enhances resale value because some home-buyers set a limit to how many steps they'll climb from street to house. Replacing a fifty foot walkway containing ten steps with one having six steps and some barely-there slopes may scoop in more potential buyers for a quicker sale. Replacing a low porch with a sloping ramp even if no one in your household uses a wheelchair—yet—can be a strong selling feature at home sale time.

Building code for steps is 11" deep treads and 7" high risers. What this means is if the tread is shorter or the rise is higher, it becomes dangerous. Which defines the 11" tread and 7" riser as right on the bitter edge of bad.

Any tread between 13" and 22" is better. Any rise between 4" and 6" is better. Any combo of the two is luxury.

A contractor will tell you he wants to do 11" tread and 7" risers. It's irrational to stick right to the edge of hazardous, within a ¼" of illegal, since sprains and falls caused by steps are a significant percentage of home accidents.

If you stick to your guns and have the space, getting a 4" rise and 20" treads will be a luxury you will appreciate every day. Or really, any combo that moves to something safer than 11" and 7". If your basement steps could be two feet longer, take it to reduce the rise or lengthen the tread. It could pay for itself in injury reduction. If you have a single step between rooms, say kitchen and living room, or one step at the patio door, never allow it to have less than a 16" tread; the single step between rooms is an ankle twister. Eighteen inch minimum is advisable.

What about the contractor who says not possible, he wants to use the stair supports from Home Depot? He's being lazy. The step cutout is a 90° angle. Anyone can buy two big boards, say 16" wide, make a template out of a sheets of legal-size paper taped together and tip it on the board until there is a 13" tread with a 6" rise, then cut with a jigsaw. It's a single hour of work for a skilled tradesman who proceeds in a workmanlike fashion and you'll enjoy it daily. Tip: he should drill a hole at each inside corner and make his cuts to the hole, not beyond.

Most state codes require a handrail when there are two or three steps in a row. Sometimes a builder circumvents this on outdoor walkways by putting a single step here and there. It may save him some money on the project, but statistics show these single steps randomly placed are a main cause of sprains and falls in all climates. When it's icy or snowing these single steps are not safer than two steps with a little handrail. Require your contractor to group the stairs when possible. The handrail is something you can buy on-line and install yourself or hire a handyman if he won't include it.

The railings around decks and high porches are different than handrails, and different sections of building code cover them. A manufacturer may make both but there are key differences. It varies slightly by state but in most cases, deck surrounds have a **minimum** height of 36" and there is no limit on the top bar width or thickness; it can be wide, suitable for setting drinks and even plates upon. Below the top bar there must be slats, pickets, balusters or panels preventing children, pets and adults from falling beyond it.

In contrast, a handrail has a **maximum** height of 36", must have a grippable rail of a maximum of 2" or 2.25" diameter, and needs only a post or mounting bracket every four feet. A handrail is measured from and follows the plane of the front edge of the steps. Aim for a 32"-33" height on handrails, which enables most adults to put weight on it without an awkward elbow angle.

Kitchens

Remodeling a kitchen with the latest trendy fad is easy to do. The stores have a lot of it on display, and they know they have to move inventory on fads because once they pass the peak, no one will want it. It's hard to tell what is a flash in the pan and what is truly an improvement that stands the test of time. Very smart people once said microwave ovens were a fad.

Things that are comfortable, convenient, and don't require frequent cleaning to look good are winners. Right now the fad is for range or cooktop surrounds that look like a temple, with pillars and plenty of scrollwork. Something like a temple look can be achieved economically with distinctive tile going to the ceiling and a metal exhaust hood.

Busy intricate woodwork is a poor choice for a kitchen, unless cleaning it will be left to your cleaning lady. Otherwise, life is too short to devote 40 hours a year to cleaning greasy niches in woodwork.

Another fad is to install a huge farmer sink, a.k.a. butler's sink. These make little sense today, when we wash all dishes in the dishwasher and the average household eats out three times a week. In fifteen years they will look as ugly as those old 1930s sinks did in the 1970s, where they still lingered in neglected houses. If you install one today, the dated, eyesore farmer sink will be the motivation to rush into another expensive kitchen remodel in fifteen years instead of holding off until twenty.

On any weekday day in the US, fewer than 1% of eat-at-home meals involve four or more people. We are a nation of people who eat meals mostly in ones and twos unless we dine out. We are a nation of people with two linear feet of cookbooks in every kitchen, but 70% of our meals involve the microwave. Never before has the myth and the reality of what we do in the kitchen been so far apart.

When considering a kitchen remodel, decide how much faking it to support the myth is worth to you. Do two-person households need 6-burner stoves? Twelve foot granite-topped islands? Forty-eight bottle wine chillers? Do you throw parties for 50 people more than four times per year? Just how many parties could you have lavishly catered if you skipped paying for an addition to add 60 square feet to the kitchen?

If your situation is that you are not cooking for nine every day, but have quite a few specialty items and dishes—a wok, two blenders, three coffee makers, five different crock pots and big cookers, twelve large holiday serving plates, three sets of silverware—then what you need is ample storage, not a huge kitchen. This need might be better served with a floor-to-ceiling 32" deep pantry closet adjacent to a smaller kitchen.

Take a hard look at your lifestyle before plunging ahead with the builder's cookie-cutter kitchen plan based on myths that don't fit you, heck, don't even fit 98% of American households. With deep thought, it may make sense to take some of the kitchen square footage and add it to the living room, or blend your kitchen/dining/living room into one, or even have the kitchen appliances line one of the living room walls, with an eat-in counter in front of them. Households of under three people might find a simple bar sink next to the dishwasher, large enough to wash vegetables, provides more counter space, which is what they really need.

Kitchens are very individual, so there is no design idea that works all the time. Scale, storage, optimal plumbing location and windows must all work together. Today, most kitchens aren't designed to be user-friendly at all; they're assembled from parts, and the guys making those parts just want to standardize their production processes. Your kitchen contractor is only going to care about making what he can buy fit into the available space. Beautiful kitchens frequently have poor workflow, awkward reaches, and are labor intensive to use and clean. If you look closely at the huge kitchens in lavish homes, you'll notice that there are basically two kitchens—two sinks, ovens on opposite ends, two garbage can locations, and so on. Only one of them is used 80% of the time. The other is for show, like an all-white parlor.

Every good kitchen must have a pet feeding location. It cannot be ignored at the design stage. Even if you don't have a pet right now, over 80% of households do.

Every kitchen must have a specific, large, convenient and esthetically pleasing garbage can place. If yours does not, the garbage can and recycle tubs will have to sit out in the open, in front of some cabinets, ruining the magazine-ready look that cost you $28,000. Worse, you might end up placing it far from the kitchen sink and far from handy, and the walking back and forth to discard things will drive you nuts.

Don't let any builder, designer, or kitchen center set you up with a design that has you desiring alterations in the first week.

Here are some additional ideas; every kitchen should scoop up a few:

Microwave: bottom floor at least 2" lower than armpit level, for better visibility and a good lifting angle to reduce spills and burns

Extra large windows, with ample sill for growing herbs.

Tile or solid backsplashes

Taking an idea from industrial design, cut a hole in the counter and install a garbage can on slides directly below. You will be the envy of all your neighbors for this alone. While it seems very risky to cut a plain hole, round or rectangular, in your countertop, you'll soon feel sorry for people without one. The trick is to ensure the garbage can (a custom-made straight-sided bin, not the taper-bottomed plastic ones) has a mouth which holds a standard 13 gallon bag snugly and rests 1" to 2" below the counter, not having a 5" or 6" gap.

Location for a monitor; most recipes are stored on a computer, or people watch TV or movies while working in the kitchen. Mounting it on a swing arm makes it convenient during food prep, and then it should tuck away nicely when not in use. A modular outlet with Ethernet, phone and coax (cable) nearby is handy.

Place for cookie sheets and cutting boards.

In climates that get below 30°, kitchen sink not on an exterior wall. The single best thing to prevent pipes freezing is to run them at least three feet away from the outer wall.

Place to set cat or dog food dishes. Not a trip hazard or in the swing range of a cabinet or refrigerator door. Not next to the stove or dishwasher.

Place for coffeemaker, toaster, etc.

Place for cookbooks.

An integrated desk area for mail, kids doing homework, etc. will be enjoyed forever. No chance of it not being used. In fact, it cannot be too big; no matter what the size it will always be too small.

Good lighting. If a one-story house, consider a solatube or competing product. Consider several of them. Each one provides far more light than the diameter would seem to allow, for half the cost of putting in another window.

Large drawers in lieu of under-counter cabinets. Never again get on your knees to get something out of a cabinet.

A great sink faucet. It tends to become a focal point, so spending an extra $100 has a huge impact on the overall impression.

A floor that works with the countertop material and color.

A pantry area or pantry cabinet. A floor-to-ceiling unit the same depth as the refrigerator can hold more stuff than four beneath-the-countertop cabinets, with better accessibility.

An opening or clear view to the living/family room. Today, with refrigeration, dishwashers, exhaust fans, central air and glass cooktops, there's no reason for the kitchen to be a separate room. It used to be a good idea due to the heat and smells, but those days are gone.

Doors

Consider pocket doors when possible. Good locations for pocket doors are master bathrooms, some closets and home offices. When the door swing would protrude into a hallway or inevitably hit furniture or fixtures, a pocket door might serve better. Pocket doors slide right into the wall, like a patio door, except when fully open the door itself is entirely hidden. The ones with smoked glass, meaning foggy glass, let in light but protect privacy.

Consider dutch doors. The half door, at first blush, seems hokey and too much like a store, but they are used to this day because they are really useful. Most dutch doors can be latched to behave as one door most of the time, and no, the little shelf atop the bottom half is not mandatory. These can be very useful on interior doors in houses with pets and children, in lieu of baby gates.

On interior doors, you will get builder's cheap unless you ask for something better. A little money gets you a huge increase in quality for interior doors. There is no compelling reason to have every interior door in the house match. On a two-story house no one would notice, maybe not even people who live there, that the downstairs doors are wood-colored and upstairs they are white and paneled, or vice versa.

Windows

Low-E coatings on windows can be a good idea, but are often done wrong. In climates where the heating bill is higher than the cooling bill, a high solar gain (as opposed to low solar gain) Low-E coating applied to the interior pane of the window is optimal. Low or medium solar gain Low-E coatings are only appropriate for windows in higher cooling cost areas. It's a five-minute read to google about Low-E windows on the internet. What pane the Low-E coating is applied to, either the interior pane or on the inside of the exterior pane, is critical for optimal performance. Most people would do best with interior pane on north-facing windows. On the other sides of the house, exterior pane for hot climates and interior for colder ones.

Unfortunately what can happen is the contractor has a one-size-fits-all type, and to make a sale he'll say sure, whatever you want. Then he installs the same thing all over. This happens often. Whatever the product, if it is hard to tell after installation whether he complied with your request, he might install what he can get cheaper or easier unless you police it before the installation starts. Write in your contract that you must be contacted to review windows in their boxes, with paperwork, prior to installation. Don't leave it unsaid what the outcome would be if you find a discrepancy; right in the same sentence in the contract specify the GC must replace them if the wrong kind are installed 'mistakenly.' If you say nothing, you have to take him to court after the windows are installed, and the judge could render a verdict for the cost difference only, not replacement.

Anti-UV coatings on windows or adding a tint is also not for all windows. Plants will not grow well beside these windows, and if you like catching some Vitamin D rays from the comfort of your own house on cold days then anti-UV windows are not desirable. Fading of furniture and so on is grossly exaggerated as a selling point; most people over thirty-five grew up in homes with plain old glass and what little fading there was took many years. The kitchen window over the sink is one location where an anti-UV coating would not be an advantage.

When replacing windows, it's possible to get a bigger or smaller one in the location. Replacing with the exact same size will cost less, but on many homes the cost to go bigger is not as bad as you think. Older homes may have non-standard windows so the cost to custom-build a single window to that size can make going larger to use a standard window a cost-competitive choice.

Bigger looks better from the outside because smaller requires patching in some new siding or bricks. The lowest-cost way to go bigger is to keep the same width and top frame but make it longer, lowering the windowsill. It's always possible to go bigger in width and height. The procedures to do this are all tried-and-true, just as they are for adding a window where none currently exist. Doing it well is a matter of conscientiously following best practices. A skilled installer can make sure only the necessary amount of siding or brick is removed. It will require drywall replacement inside.

Going bigger on a window or putting one in a new location is not a DIY or first-timer job. It's important that your installer, this particular person, has reframed a window several times before, not someone else at the company that employs him. If the hardware store you bought the window from says they have done 20 of these in the past year but they send a 25 year old installer they hired three months ago, send him back.

Drywall vs. Plaster

Occasionally you'll read that someone thinks plaster is more high class or better. He or she's an idiot. Plaster walls are more labor intensive to put up, but higher cost doesn't mean better.

Plaster walls are heavy. They're prone to crack when normal house settling occurs over the first few years. They suffer water damage far worse than drywall, which might discolor but won't come crashing down into the room after a few hours of soaking. When surface damage occurs, it's harder to repair plaster to invisibility than drywall. Drywall harbors wall hooks well; plaster can chip out when pounding a nail. When affixing heavy items to the walls most people can make do using the studs on 16" centers behind the drywall; with plaster, most people depend upon the weaker laths to hold it up. Finding studs under plaster walls can be difficult to impossible with the usual studfinders.

Later on when remodeling, it's a nightmare to remove plaster and lath walls compared to the minutes a drywall wall requires. Plaster dust can ruin the TV, phones, cell phones in your pocket, and other electronics anywhere in the house if you don't plastic bag them all. If you're stuck with plaster walls you can do what the Victorians did; cover it with busy-patterned wallpaper. Wallpaper not only covers the cracks, it minimizes the chipping out around picture hooks and slows the progressive damage of humidity.

Drywall is inherently fire retarding. Mold-resistant coatings are available. Sound-proofing, considered an advantage of plaster, can be obtained with insulating or sound-proofing materials installed during construction. As an aside, all advantage of sound-proofing the walls is lost if adjacent rooms share air vents or air-return ducts, which are metal-lined paths that bounce sound amazingly well.

Roofs

A whole book could be written about what is wrong with roofs. To be brief, I'll point out that in Europe there are millions of homes whose roofs last for 100, 200 even 300 years with only occasional repairs. Here, we can hardly muddle past 25 years without total deterioration of our 'superior' and 'engineered' materials.

The science and technology are there to make roofs that will last for 100 years, and if we all bought them the cost would come down to just about the cost of a new roof today. We don't have that option because builders install what they like, and they like frequent roof failure. Today there are adhesives that will last for 50 years and not be affected by water and sun. Why we are still nailing?

So basically I'm saying all over Europe they've had the art of making sturdy roofs mastered for several hundred years, and today we have advanced technology at our fingertips to do it even better, but what we actually install in new construction creates inevitable roof, wall and window failure between 15 to 40 years.

Thinking about that doesn't help you get a good roof today, but there are a few things you can demand of your designer and builder that will improve enjoyment and reduce repairs for the next 100 years.

Insufficient roof overhang and lack of a roofed section over doors cause windows and door frames to have water-related issues in twenty years, while houses with roofs extending a foot past the walls and with spits of roof over the doors and lower windows will commonly see frames last eighty years and more. A roofed porch over every entrance not only looks better, it is user-friendly, protecting the owner from the elements while he or she works the locks, and protecting visitors while they wait for someone to answer the doorbell.

Rainwater sheeting down the exterior side walls until it encounters a window has a way of wearing down caulking and getting into the walls. Overhangs and decorative raised framing over windows reduce the sheeting action on window and door frames to almost nothing. A wonderful side effect of a decent roof overhang is you can sleep through rainstorms without rain hitting the window and sounding like a drum.

1. Sufficient roof overhang and soffits on both houses; good design

2. Barely enough – the minimum. No room for soffits.

3. Ridiculously inadequate roof overhang; shows evidence of replaced siding by windows due to rot; all three houses are about 25 years old.

Recessed windows under roof and integrated porch roof prevents wood rot

A good overhang that protects windows and front door

Older homes in Italy with typical generous roof overhangs, large and plentiful windows that are recessed so rain never hits the top of the window frame, plus sheeting-rain diverters are stylishly built-in.

Typical terrible US design. Neither arched window nor door has rain protection. Dinky front porch will not hold three people. Roof overhangs are too small to harbor sufficient soffits, so the builder popped a round hole in the siding—too far from the peak to be protected by the inadequate overhang. No gutters, so rainwater cascades down to the foundations. In a heavy rain, one roof angle will spout its full load right onto the front porch—do you see it?

Plywood, Particle Board, and OSB

One is good and durable, one is for temporary use only, and one is dangerous.

Plywood is what you want in your home, and plywood is what you must compel your contractor to use on your projects big and small.

Particle board is made up of sawdust-size particles glued together. It's heavy. It's the wood product used in low cost assemble-at-home furniture and shelving.

OSB stands for Oriented Strand Board. OSB is made of bigger chunks of wood, also glued together. The difference in cost between plywood and OSB is about $600 on a 1800 square foot house, peanuts, yet a builder will ruin your life and your financial future for a measly $600 if you don't fight it like the dickens.

OSB expands like a sponge when wet. Unless your house or addition is completely finished in several non-rainy days, any OSB in the construction area will get wet. That water will be retained as the walls are closed and flooring is installed. In a few years the mold problem will cost you tens of thousands of dollars to remedy.

The glues used in OSB turn out to be a great food for mold. If you're lucky enough to evade moisture exposure during the build, when water infiltrates at any time in the future, from an overflowing sink to a broken gutter, your mold problem will start at that point.

If mold isn't a strong enough reason, then forbid it on your construction because it swells and warps with mere ambient air moisture. Tile floors and walls should never be installed over OSB.

Another reason to shun OSB is because it doesn't hold nails even half as well as plywood. When OSB is used under roofing and under siding, in high winds there goes your shingles and siding panels. It contributes to all kinds of squeaks and nail-popping-out issues throughout the house. When stressed at the nails, whole chunks can loosen and chip off, meaning then it isn't even a matter of putting a screw where the nail used to be, like you can do with plywood. There simply will be nothing touching the screw threads.

There are many online sites that say this stuff is great, stronger, better, yada-yada. They are all trying to sell it or are written by people who drank the Kool-aid. As the amount of failing homes grows, the people who parroted that baloney about OSB are going to slink away in shame.

It may be possible to 'fix' OSB –stronger and mold-resistant glues, an OSB-plywood layered sheet—but at this point the only sensible thing is to demand plywood in all your construction until any purported better OSB has been out there for ten years and the marketing hype is panning out, say sometime after 2025.

Great Home Features for Pennies

View of OSB, from Wikipedia

Expensive benefits that are not: Today's trendy thing can seem so compelling that contractors believe, without lying, that it's how things will be done from now on and it isn't about taste or style. Trendy things reveal themselves only in the rear view mirror.

If you're young you have no ammunition against this; older people can remember when shag carpets, avocado appliances, pink toilets, carpeting for wainscoting, popcorn ceilings and pedestal sinks were top choices and how compelling they were at the time. The contractor's favorite argument is that not doing it will affect home values. Listen to him if you're sixty-five and want to do it the old way; if you're thirty and don't like it, trust your gut.

Old trends may look funny, but there are plenty of innovations that stand the test of time: walk-in closets, 3-way light switches, under-the-cabinet mounted radios, patio doors, built-in cooktops, powder rooms, central air, skinny cookie sheet cabinets.

New trends that may stand the test of time include: big drawers and no cabinets in kitchens; feng shui for entrances; bottom-drawer freezer sections on the frig; wall-mount TVs; built-in desk in the kitchen; huge wall mirror above the bathroom sink; outdoor kitchen; patio door & deck off a second floor bedroom; door-free bathroom showers; in-floor radiant heating for sunrooms and rooms where air vents would ruin the look, such as rooms with curved walls.

A contractor might be enamored with certain features and try to sway you with suggestions. Say he's into ceiling treatments, fancy cornices or tin ceiling plates. He may suggest doing a tray ceiling to lend an appearance of height to the ordinary eight foot tall ceilings.

After he's done not only have you paid more, but the French flourishes clash with your Early American style—wooden beams would have looked better—and the tray ceiling just looks silly in such a small room.

Replace 'ceiling' with any other part of the project; things may be touted as 'better' when they are merely in style. It's funny how popular opinion can cool on previously ubiquitous choices.

Sometimes a more expensive product is pushed, such as in the countertop arena, for benefits that may not matter. Other materials can be pegged with a negative for something that happens rarely.

One example is solid surface countertop material that looks beautiful but could be damaged by hot pots. The pitch for higher priced granite is that pots right off the stove won't damage it. Big selling point. Anyone raised in homes with laminate countertops tend to be extremely well-trained to never touch a hot pot to a countertop. Nice feature, but not worth a dollar to me.

I've never put a pot burn on a countertop in my adult life. There's no doubt on that point because the burn would remind me for years. High priced countertops are pushed by sellers because they are high priced; consider the least-costly one that looks good in your kitchen.

Ergonomic improvements

It is worthwhile to overcome initial contractor resistance to making small upgrades or ergonomic improvements that he didn't think of first. In some cases after doing it for you, he will warm up to the benefit and make it part of his offerings, becoming a competitive advantage for him.

Don't get disheartened over initial contractor disapproval if it's something you think would be convenient or helpful to you. Give some thought to his logic, but make sure it's logic. If his reasons amount to 'I haven't done it this way before' or 'it's more work,' it means there is no reason at all. He's only being negative.

Schadenfreude

Schadenfreude is a German word meaning taking pleasure from the misfortune of others. We've all heard dozens of contractor stories, and all the interesting ones are bad. It's easy to find tons of long stories about failed projects. They all boil down to four things: one, simply being what happens when a first-timer at this particular work is hired; two, a green crew was assigned to the job; three, a swindler was hired, one who never intended to complete the work at all.

And four, the remainder entail running into structural defects, mold or insect damage, which would have happened no matter who was hired. The details are immaterial; it's one of these four.

Reading about horror stories will not help anyone do a better job.

Good contractor experiences proceed like this: I hired a contractor, he came on time, he did a good job in a timely fashion, he didn't break anything, he gave me a bill for 90% of the contract price when he was done, and twenty days later I sent him the remaining 10% because nothing was wrong. Yawn. How many times do you want to read that?

You can have that same boring life. In the realm of motivational spurs, however, boredom is not a good carrot.

Psychology has been studying how people learn for decades, and the jury is in: scaring people with personal injury sucks as a teaching method. It creates avoidance, not a desire to adopt best practices with a gusto. In the realm of home improvement, overdosing on bad stories will subliminally lead to you to minimize contact, minimize research, and feel overwhelmed. Enter your contractor situation believing, "Say, I can do this; it isn't that hard; half the work can be done while watching NCIS."

Train wrecks make fascinating vignettes, but studying eighty projects gone off the rails teaches you how to do it wrong, not right.

When your parents taught you how to drive, as you approached a 4-way stop, did they describe severed limbs? When they showed you how to use the turn signal, did they hold up photos of bleeding head wounds? When you entered a circular freeway ramp, did they never fail to tell you about the poor fellow who became a quadriplegic when, going too fast, his car slid off the ramp and rolled over? OK, I confess I did that last one when teaching my son how to drive.

Your parents weren't being naïve and weren't withholding information when they put aside, for now, the possible consequences of not coming to a full stop or not using a turn signal. Time has proven that simply teaching someone how to do it right is sufficient. Learning a new skill while being terrified of the maiming and disaster that awaits like a dragon at each intersection neither kindles confidence nor bolsters the learning process.

In short, it doesn't help at all.

Learning the right way to select, hire, pay, oversee, and manage contractors is just about the same as learning to drive. When learning to drive, at first it feels like a mishmash of unrelated details. Several things are all happening at once, or one immediately after the other. Feet, hands, eyes all have to do their own thing concurrently. In just a few hours of practice, the motions become natural, routine and obvious. As tense and impossibly complex as it feels at first, it very quickly becomes routine. The voice in your ear telling you the moment to flip the blinker becomes a motion at the proper time without thought.

You can hire a contractor and have it turn out well. Just follow the rules of the road, steer clear of suspicious drivers, and you'll get to your destination without bad things happening. If you are so scared of the bad things that you decide to minimize your exposure by racing as fast as you can while not looking sideways, then most likely bad things will happen to you.

There is another similarity between driving and being a good contractor-hiring person: there are many things you have to do 'just so.' Everyone learns to do them; it doesn't entail a personality change. Within the full range of good drivers, there are timid people and bold people, strong and weak, tall and short, calm and nervous, happy and sad, quiet and talkative. Meaning when learning to, say slam on the brakes hard, timid people will need to work on it more than angry people. When waiting your turn at a 4-way stop, mild, gentle people will have an easier time than hotheads. But in the aggregate, all of them can do it well enough. Good enough is truly good enough. I am not perfect at my own advice either.

This chapter contains some real stories meaningful to my own learning curve. I hope they help or amuse you.

Mike and the Black Walnut Tree

Years ago a co-worker of mine, 'Mike,' needed a tree removed from his yard prior to adding a garage. As co-workers do, he was venting to a few of us on what high quotes he was getting. He had nine different tree-removal firms come for a site survey of the tree on his property, and the quotes ranged from $800 to $1200 to cut down and haul away just one tree, and that didn't include stump removal. Another co-worker, Leo, thought that was high too, and asked how big it was, then what kind of tree it was. Mike replied, "It's over two feet in diameter, a Black Walnut." Leo looked at Mike quietly for a few seconds, and then said "If it's a Black Walnut, they should be paying YOU, not you paying them." A Black Walnut tree of that size sold for about $2,000 back then.

Armed with his new information, Mike made some more calls for quotes. The offer accepted was from a shop that paid him $400 plus chipped out the stump for free.

The tree didn't change, Mike's address didn't change, and the amount of work didn't change. What changed was Mike's expectation of the price.

In the earlier round of bids he telegraphed, or perhaps stated outright, that he expected to be paying them. No one disabused him of that misconception. No one gave him even a tiny break on the usual tree removal price. With the second round of bids he showed he knew, resulting in no possibility they'd go in the wrong direction with their bid.

Were the first nine guys all swindlers? No, they were reputable businessmen being opportunists.[23] It's likely they do a perfectly fine job of removing trees. Had Mike called them with the right price in mind in the first place, one of them might have offered $500 and stump removal and gotten the job.

There are estimators who price by cost charts and don't consider customer gullibility in their quote. But as Mike's experience with nine different vendors shows, you can't count on bumping into one of those guys.

The nine shops weren't in cahoots, phoning each other in the evening to say "Hey, Mike might call you tomorrow, don't let on how much a Black Walnut is worth, be sure you charge him the full amount of a tree removal."

No, the identical thought dawned on nine different businessmen. And if Mike called another six tree removal shops asking them to beat the low bid for tree removal, odds are all six would have done what the first nine did. Charged him.

This is an excellent example for another reason. It might seem plausible that overcharging would be only a few hundred dollars or perhaps within a percentage such as 10% or 20% of the real price, hardly worth expending any effort researching pricing on jobs under $1,000. Not true.

It's not just the big jobs and there really is no ceiling on how big the rip-off can be. A friend of mine paid $900 for a 2-hour paint touchup,

---

[23] Ironically, had one of the first nine quoted a price of $400, that guy would have gotten the job and Mike would not have mentioned it at work. Their greed for a few hundred more meant they lost an easy $2,000.

materials under $30; $250 would have been a reasonable price. A little old neighbor lady received a $4,000 estimate for chimney repointing, which for her chimney was a one person, one-day job with materials costing about $50, meaning the only way it made sense was if labor was $500 an hour.

Playing My Side

On my second house I hired a contractor to install a patio of stamped concrete in the back yard. The house came with a cheap deck in bad shape in that location. After agreeing upon a price for the work and picking my pattern and colors, it dawned on me that we'd been silent upon the old deck. I called the contractor and unabashedly asked him to remove the old deck and dispose of it too. Period. Hardly skipping a beat, he cheerfully said OK. Conversation done. The old deck was removed for free.

That little victory delighted me for a few minutes. Of course the only thing that made sense was that he'd assumed it was part of the job and had already priced it in.

Making the same phone call but asking for the price of the work or saying it is a change notice means he would have naturally agreed and increased the price. Price increases are never something a customer requests; always assume the current price is for all the work, even work you didn't mention yet. Never volunteer to pay extra; his job is to charge correctly for his work, your job is to keep the price to a minimum.

Uncertainty = more money

At the early stage of a large grounds project, after conducting the walk-arounds with each of the contractors being considered for the work, the quotes came in. The bid of a preferred vendor known for doing great work was a bit high. Instead of asking for a lower price, I phoned to ask if we could have another brief walk-around, saying we could view some of the trouble areas again. He eagerly accepted. The following day he faxed a quote several thousand lower.

Why did this work better than simply asking for a lower price? Typically, when sitting at his desk several hours later jogging up the quote, his skimpy notes aren't enough help or he forgot to take a measurement, so in the spirit of completing it by deadline he'll assume a worst-case situation. Sure, he could drive back and trespass to peek again, but that could become awkward if spotted since his motions could appear very suspicious.

With the second walk-around he reassessed risk and adjusted several worst-case numbers downward, providing a lower bid. If the project is large and you suspect the price is inflated with too much risk money, the best approach is to get that risk money out of there.

One-Bid Opportunism

During the first months of my term on a condo board we needed to replace some old paver sidewalks on forty homes. I was willing to handle it, but it seemed unlikely to the rest of the Board that I just so happened to be qualified to do this, so the initial plan was to hire a professional in the concrete business to help us pick a contractor. A bid for being our middleman was solicited from only one contractor.

The tasks we asked of him included buffing up my RFQ, which I'd already generated (while watching TV), distribute the RFQ to sidewalk contractors and conduct the walk-arounds of the property prior to bidding. Then collect the bids, review them and recommend one. The idea was that having a concrete professional as our middleman would get respect and guarantee fair prices.

He presented a contract for $16,000 to perform tasks amounting to sixty man-hours of work and no material expense, other than some paper and postage, meaning he was hired for $260 an hour. Plus, he required over 60% be paid by the first week of a six-week contract. The Board, consisting of intelligent college graduates who were totally out of their element, voted to accept the deal. While the blatant opportunism stuck in my craw, I begrudgingly agreed that if it helped us avoid the much larger pending opportunism, it might be forgivable. Unfortunately, the paying-too-much aspect caused disrespect and insolence that bit us on the performance side. The experienced owner handed off most of the work to his young son, who didn't have the skills to comparison shop for a TV.

The result of this amateur help was nine quotes, not one covering all of the requirements in the RFQ and all except the most expensive one omitting both handrails and reseeding the lawn. The initial quotes ranged between $250,000 and $438,000 and all were missing major RFQ requirements. The middleman, now fully paid, refused to recommend one, and refused to call them back to get the missing aspects included, i.e., make the quotes to apples-to-apples instead of apples-to-oranges.

The Board itself had to go over the bids line-by-line with each of the bidders. It turned out all his bidders quoted grey concrete, not the finish we wanted. Every one said the price would increase somewhere between 15% to 50% for colored concrete with a decorative surface. Some flatly refused to quote handrails. Some of them had working timeframes over twenty-five weeks.

These prices were too high by a mile. The twenty-five week timeframe was merely the common tactic of stretching out the duration to justify the super-high price tag. It was not possible to be a twenty-five week job with workers here every day. To explain that, concrete work is on the clock; too much space between one step and another and it has to be done over.[24]

After some discussion of these details, the Board reluctantly agreed with me that these were ruined vendors—they couldn't cut their quotes in half without admitting how seriously they intended to rip us off. We had to start over. We said thank-you-goodbye to the cement guy middleman. Deflated at having spent $16,000 and losing two months for absolutely nothing, the Board let me take over. I cast a wider net for fresh-meat sidewalk contractors and sent out another round of RFQs.

After finding four new contractors and performing the walk-arounds myself, we had an excellent batch of quotes. On the walk-arounds I listened, asked questions, shared the best ideas with all, and thought ahead on the risk factors. The result: four quotes for exposed aggregate colored concrete that looked like a winding pebbled path through our wooded grounds. Three of the four were under $190,000 *including* handrails and reseeding. Pricing on the middleman's batch varied 75%; my batch varied 4%. Today, the sidewalks, porches and handrails are lovely. The project came in on time and on budget, except for extras we needed such as some asphalt patching.

Adjusting for the missing handrails, seeding, sidewalk coloring and custom finish in the first batch, the accepted quote was half of the very best price the concrete professional obtained.

Could some of his batch have been highballing because they didn't want the job? Possibly. Were some of them first-timers at sidewalk work? That could explain a high price; paying for their learning curve. Were the guys in the first round of quotes all pulling a fast one, and in the second round were honest businessmen? Not a chance; both rounds were just local businesses, all of them well-established. As you might guess, the closest businesses and the ones best-sized for our work were in the first batch.

The second batch prices were consistent and low due to my work with the estimator on the walk-arounds. Watching Youtube videos on cement pouring and reading DIY sites on the steps for prepping and pouring a decorative sidewalk helped with terminology and understanding the job challenges. For instance, we had to talk about shrub handling (marking which stay and which go) and accidentally-cut Cable TV lines (agreed to

---

[24] This job would be done using full cement trucks, nine cubic yards of concrete, so that's the amount they would prep and pour each time. The length of the sidewalks times width times 4" thick permits calculating the cubic yards of concrete. Mentally grouping the sidewalks per truck leads to the amount of pours. In our case, nine trucks. Underlayment, compacting and framing should sit less than a week before a pour. So, roughly one pour per week dependent on weather. Days that are too hot, too rainy and too cold hamper both prep work and pours, so throw in another three weeks for lost days. Total: about twelve weeks. Then another two weeks for fixing lawns and installing about twenty hand rails beside steps. Although these could be concurrent to most of the project, they probably would be done as a batch at the end.

phone the Condo Manager who would make the necessary arrangements with Comcast). The average sidewalk was emphasized instead of the worst ones. I reminded each estimator several times that twenty-eight of the forty sidewalks averaged twenty four feet long. Complexity was downplayed by mentioning the one- or two-step porches were simply thicker sidewalks with framing that could be reused several times.

A few homes had several steps and some long winding paths around landscaping. By counting them ahead of time and saying we need 99 steps, all of them alike, it turned them into simple math. One beginner mistake our middleman may have made is to have spent long portions of the walk-around going over the long walkways with a lot of gesturing and problem-solving while breezing by the short ones, so the mental picture retained when jogging up the estimate back at the office was of the worst ones. During my walk-arounds we parked ourselves in front of some short ones while we discussed other details, and used the short ones to illustrate points like removal and porches.

Best practices that reduced logistics[25] risk and access issues were shared with all. In the few areas where access was tricky, everyone got the same information on how they could bring their trucks close or reach it better from the back yard.

Because I'm an old pro at this, I suggested an estimating approach, recommending they price the short walks and add on for longer ones rather than price long ones and subtract for shorter. It sounds like semantics, but it works out differently.

Optimizing the design to the materials was permitted. For the houses dropping off steeply in the back that had staggered four foot high wooden retaining walls holding up a pile of gravel for the porch to sit on, they were permitted to optimize the design for solid concrete rather than imitate the limitations of wood. Areas of cost reduction were mentioned. Since the ground under the current sidewalks was well-packed down, less excavation and fewer compaction passes would be needed than on new construction.

The lowest price doesn't happen by having a job title in the trade; it happens by sharing information, addressing contractor risk, wording things to reduce complexity, exhibiting ethical and honest behavior throughout, and allowing them to use their expertise to be efficient.

After we hired a contractor, watching the work and speaking up when needed was vitally important to a good outcome. I was present all day for the first few pours, checked the framing before every pour, poked at the compacting a hundred times and asked for more whenever it sat for a few days, used my frowny face as needed, reminded about shrub handling and

---

[25] For an explanation, see Addendum B, The Ten Risk Factors

sewer lines as often as I wanted to, reminded them to call the Manager as soon as a wire was cut, no saving that call for later, and so on.

On the second pour, the cement truck was here for several minutes before someone hopped up on the side of the truck to the tumbler opening and began adding the color. I advised the Foreman that from now on, when the truck arrives let's see someone hopping up there with color bags in hand within fifteen seconds. After that, it happened. Yes, the project went well. It went well with me as a full partner in the good outcome. Shrub handling and porch layout errors were reduced by my involvement. My contractor was his best self on my project. At the end of the day, that's all I want, a good outcome.

Keep the Verbals

One entrepreneur speaking at a New England Inventor's Association (InventNE.com) meeting shared several of his startup troubles with the group. When the design was satisfactory for his plastic part, he went hunting for a manufacturer to build his first large order. During a tour at a plastics manufacturer, he observed an operator packing finished product into cardboard boxes with dividers, one part in each compartment. That's how we pack all our parts, the estimator told him.

He signed a brief contract for an order of several thousand parts. Weeks later, his order arrived. The thousands of parts were simply tossed into two large boxes, which were roughly handled and plopped on his driveway. Over 90% of the parts were broken. The contract they signed contained nothing about the packaging or shipping method. The manufacturer denied agreeing to any other packaging method. In the end, the entrepreneur was simply out the money. While he had actually requested to be phoned upon completion and intended to pick up the parts himself, he only asked for that verbally and did not bother to add it to the contract.

It's a good story showing that even if the estimator says 'we always do this' or it seems to be common sense, if you really want it to happen, it needs to be stated in the contract.

Hydraulics

When I was 26 I decided to tackle replacing my bathroom faucet myself. In college I earned an A in hydraulics class plus had assembled several high-pressure hydraulics systems on fixturing at work so working on low-pressure household plumbing should be a piece of cake, right? My work buddies laughed and shook their heads. One or two offered to come over to help me out. I refused those offers, but listened to their advice on the tools I would need. On Monday at the coffee break my buddies asked me how it went. "I did it like a man." I replied, "It took four hours longer than it should have and I went back to the hardware store three times."

My advice is once you embark, do it like a man: give it all the time it needs, don't give up too easily, and find the most knowledgeable friends and hardware store staff and tap them for their advice. Avoid working on a Sunday; the stores aren't open as late and if something goes amiss getting a real tradesman to rush over is expensive.

Roofing

Over ten years ago a condo board hired a roofing contractor based on the gushing recommendation of one of the members and an impression that no other contractor in the area would do projects of that size. While roofing 90 houses on one street, half of them connected in pairs, should have qualified for a huge price break, the price per house ended up being over $13,000, twice what it should have been.

There was an inexplicable amount of trust extended to this contractor, so the contract signed provided little detail. The Board signed the paperwork the contractor gave them. They paid attention to insurance and licensing, but not to the actual details of the contract itself. They did not insist upon a holdback and did not make the RFQ a contract addendum.

When the work began, the contractor tore off the old gutters and downspouts as well as the old roofing. After doing several houses, some of the owners noticed he wasn't installing new flashing around chimneys and skylights, even if it was obviously damaged. They reported that to the Board, who naturally went to the roofer to ask, "what's going on?"

With work still proceeding at a clip, the roofer said that flashing and gutters weren't included in the deal. The Board had just assumed they were, without any mention in the contract. Not only that, once they realized new gutters weren't included, the Board asked for them to be carefully removed and set aside so those same gutters and downspouts could be reinstalled. The roofer refused, saying his contract was for crowbar-style teardown and that's what he would do.

In the end, the Board had to negotiate a second contract with him for thousands of dollars more to do the flashing work around skylights, chimneys and stacks. They were now over a barrel and had to pay the price this contractor demanded. They could, however, get competitive quotes for installing new gutters, so that is what they did, for tens of thousands more.

It would be nice to think this roofer lived happily ever after, wintering on his yacht in the Keys, but in fact he went bankrupt two years later. The big selling point, a ten year free-repairs warranty on the work, went with it.

Ethics and the Best Price

When my son was 12 he liked to play an on-line interactive game called Stronghold Crusader, a world based on medieval times. One day he shouted 'Uggh!' and on-screen start fleeing another character who was yelling angrily and chasing him. This was not game anger.

"Who is that?" I asked. "I dunno" he said sheepishly. "What do you mean you don't know, he obviously knows you." Eventually I dragged the story out of him.

To buy good gear, players need to earn money. Experienced players can employ others; the worker gives up part of their earnings in return for learning the skill. Brand new players are unaware of the inflation that has overtaken this world. Several opportunist, experienced players hang around where the new players enter and offer them a wage that is 5% of the actual sale price of goods they produce, but by the booklet that comes with the game looks like 50%. Sooner or later, often after working for many hours, they find out they can't buy anything with the pittance they're paid. They quit and are pretty irate. The angry guy was someone who worked for my son weeks ago and was still angry.

I gave my son The Talk. Taking advantage of people doesn't work in the long run. You think you can get away with it, but as you learned, that is seldom the case. Now in this computer world you are always looking over your shoulder for people who hate you. Is that how you want it to be, other people wearing the white hat and you wearing the black hat? You will have the best life if you treat people fairly even if it seems at the moment you could cheat them and get away with it. If you mistreat people and take advantage of innocence you diminish your soul." OK, that last part about the diminished soul was probably not very compelling with a 12 year old.

Flash forward several weeks. My son is chortling in delight at the computer. Walking over, I ask him what's up. He replied "Mom, I have 28 people working for me."

I asked how that could happen. He replied, "Well, I still hunt for newbies to work, but I offer 25% of the sale price. I get them all because that's like five times more than anyone else will pay. Later they find out how much stuff costs and they're mad, but not that mad. Sometimes they keep working because they almost have enough for some good gear. Sometimes they leave but then come back later when they find out how hard it is to earn money."

A long pause. Then he added "When they see me they say hi."

If there was something I could say to every businessman, I'd say fair pricing to every customer big or small will bless your soul, and there might be a little earthly reward in it too.

Definitions

Air time – Between job start and job end, hours when no progress is made during daytime work hours because workers are off-site: weather-related, simply absent, or assigned to other projects. Some air time may improve quality, such as allowing adhesives to cure or cement to set, but air time lasting more than 30% of the duration on projects taking over two weeks gets into poor service territory. Air time is minimized via good planning and proper staging of resources needed for the job. Customers never pay for air time under any circumstances.

Bread-and-butter work—Work that a company or individual has extensive experience at so can do well and easily. Does not refer to easy work per se, but to high levels of experience. A tile layer who is an artist at mosaics would have mosaics as his bread-and-butter work. One carpenter might have deckwork and another carpenter have built-in bookcases as the bread-and-butter work.

Bad Karma—Behaving in ways ostensibly to gain advantage, but in the miasma of the world, will usually result in somehow, eventually, being worse for you. The seller has all kinds of ways to earn this, but as a buyer, avoid these bad karma sources: attempting to get an employee fired for breaking something or making a mistake; resorting to shady threats or unethical behavior yourself when trying to get money back from unethical guys; asking your favored contractor to match the lowball price of a creep you would never hire.

Benchmark—It means comparing a product, project or process to one or more similar ones, with the purpose of seeing where it ranks or stands compared with the others. Benchmarking is often used to determine where a product needs to significantly improve. Consumer Reports benchmarks products.

Best Practices—Combining and incorporating all the efficiencies, techniques, tricks of the trade and good approaches that can be found. Before and after a contractor is selected the project can be amended and revised to ensure best practices are employed during the project for the best outcome. Hunt for them on-line, ask experts, or cherry-pick from all the bidders.

Building Inspector—A contractor pulling a permit from your local government is how the Building Inspector or BI is notified of the location and scope of the work. He will schedule one or more inspection visits depending upon the nature of the work for the purpose of enforcing compliance with building codes. He usually has an architecture, civil engineering or tradesman background.

Callback—Phoning for another visit to fix, tweak in or improve something related to the job prior to paying the holdback. Callbacks after the holdback is paid fall into the realm of warranty calls. A callback is a normal part of most projects.

Carrot and Stick—From an old tale about keeping a donkey moving by dangling a carrot on a string in front of him, which he pursues. Or he could be kept moving by striking him with a stick from behind. Use creativity to employ 'carrot' motivation methods instead of always resorting to stick.

Catastrophic failure—An engineering term meaning a sudden, defect-caused loss of functionality or a key component ceases to work, as opposed to failure caused by gradual wear or aging. It does not always entail loud noises or falling down, but those things might be part of it.

Cherry-pick—Selecting specific bits or elements based on their appeal to you. In a legal sense it means ignoring evidence that disagrees with your position, but in this book it means selecting the best and leaving the rest behind.

Completion payment—the payment made upon completion of the job, including cleanup. This is not payment in full; 10% is withheld as a holdback.

Continuous Improvement—it means changing your mind for the better. A continuous improvement mindset is the willingness to leave your first thought or plan behind when you hear of something incrementally better, as often as several times a day. It does not need to be 30% better to instigate change, just 3% better. For instance, 'better' may mean 'easier to repair twenty years from now' with no immediate benefit, but little additional cost either.

Contractor—The hired half, with or without a paper contract. Encompasses manufacturer, subcontractor, installer, supplier, vendor. Contractors will purchase materials from their suppliers and may subcontract out parts of the job.

Debit card—A cash cow for banks and a fist-sized hole in your accounts where anyone so inclined can grab out all your money. The average debit card rings up $200 in overdraft fees annually. Unauthorized use of a credit card makes you liable for $50 max; unauthorized use of a debit card does not appear to be a crime in the eyes of the police or the bank. Even if you play detective and find the perpetrator yourself, you'll find it difficult to interest the authorities in treating it like a felony. Too bad, so sad.

DIY— Do It Yourself.

Duration—the calendar time from start of work to final cleanup and completion sign-off. Does not include the holdback period. On one to three day projects it's reasonable to expect the charged work hours to add up to the duration, using 8 hours per day. Well-run longer projects will have a duration less than 20% longer than the added-up work hours. Never pay hourly charges for hours or days when no workers are laboring on your behalf.

Final payment— See Completion payment and Holdback.

Frowny face— Making one's face consistent with the message. To not smile, grimace, or soften the blow. Put on the frowny face three seconds before expressing displeasure. That way you can give it your full attention prior to saying any words. This lets the other guy know what's coming. If you can't do a good frowny face, focus on pulling your eyebrows down and scrunching your lips together in a bunch in the middle, as if holding a straw. It suffices.

Full Burden Labor cost—The full burden hourly charge is the employee's hourly wage plus a sliver for all business expenses. Some include the profit margin in the burden, while others apply the margin last, to the whole job including materials. Businesses find it easiest to assign business costs proportionally by adding annual costs and dividing by anticipated billable hours, revisiting the Full Burden hourly charge only once or twice a year.

GC – General Contractor. This is the contractor who hires, coordinates and supervises all the other contractors on a project. He may have a crew doing part of the work, or he may only manage the project. Experience is vital to do a good job. Generally, the GC provides the warranty for the full project, providing the customer a single point of contact regardless of the issue.

Highballing—providing a quote that is exorbitantly high rather than decline the work, enabling the customer to decline the vendor. Signs of highballing are: low detail, lack of pressure, and quick response. A highballer will avoid you and act distant if you phone him with questions.

Holdback— Usually 10% for thirty days, but can be 10% or 15% for 90 or 180 days of the agreed-upon price on jobs with a high risk of issues. On large projects where seasonal settling/cracking may be an issue, 10% for six months may be appropriate. The holdback provides incentive to address issues that arise. Regardless of stated duration, the customer pays the holdback only after all customer-reported issues, punch list items and callbacks are completed. If a contractor does not respond quickly to callbacks, the customer can and should use the holdback funds to secure another vendor to do the work.

Jargon— the words and phrases that have become common terminology in a trade or industry. They will often be common words repurposed to a very specific meaning in that trade. Miscommunication can occur when a customer misinterprets jargon or the contractor thinks the customer is using a term but it was just inadvertent word choice.

Idle time — The time when workers could be working but are not due to factors out of their control, such lack of materials, missing equipment, tool breakage or a customer-generated work stoppage. Occasionally there is idle time during curing, setting or cooling; good planning can sometimes nest a useful task into the idle time. Long coffee breaks and other employee-generated non-work times are called screwing off, not idle time.

Milestone— A measurable or obvious point in the progress from start to finish. Calendar dates or payments may be keyed to milestones, which is a useful way to keep projects progressing. Milestones must accurately reflect the percentage of job completed as well as a visible accomplishment. Since any lull in work will happen immediately after a milestone payment, 'teardown complete' shall never be a milestone.

Mulligan – a golf term meaning allow the swing to be done over as if it never happened, a do-over. In negotiating it is one of your best tools to gain a better price. In one way or another telling the guy oops, that price is out of bounds, so 1) let's start over; 2) I didn't catch what you said; 3) it must be a clerical error or typo. Your attitude is, let's call that a mulligan to keep you in the game. It avoids a head-on confrontation; it permits him to change his price 50%, even 80% yet allows him to keep his self-esteem intact.

Opportunist – business that takes advantage by overcharging when he perceives the customer won't protest, can't tell, or has a time crunch. Every business is opportunist at one time or another.

Price wish— a customer expressing a price expectation so low that the estimator knows it is not based on any knowledge of material prices or labor time. A price wish damages the customer's negotiating position because the savvy estimator categorizes him as a know-nothing customer. Now he can confidently make up bogus terms and pretend issues to bolster his negotiations. With fair and honest tradesmen, pressing a price wish that removes their wage from the picture pisses them off.

Profit Margin—same as the margin or mark-up. In the estimating world it is a percentage of the actual expenses, not the year-end profit which is lucky to be 1/3 of the assigned percentage. The mark-up encompasses small and one-time expenses not accounted for elsewhere in the bid, and often becomes the owner's pay. Most home improvement/repair shops insert the mark-up into the hourly rate to simplify the math when estimating jobs.

Punch list— a written or oral list of end-of-job items the customer requests; the job is not done/completion payment due until the punch list is completed.

Risk Factor – Possibility of the occurrence of added time or lost money, the latter either through additional expenditure or not getting paid. It is determined at the estimating stage based on this estimator's gut feel plus experience. Keeping the risk factor charges to a minimum is the way a customer can keep costs down. For a full list of the risk factors, see Addendum B.

Sales Manager – many companies provide all their sales staff with this manager title; some reserve it for a handful. Its sole purpose is to render an impression to customers that his skill level is higher. It's actually irrelevant; he has a boss, and that boss will be invoked if it helps him wrest more money or better terms from you. Whether the Sales Manager selling to you manages people or not is also irrelevant; he's handling your sale.

Sharpen his pencil – It means the seller will recalculate the quote to a better number for the customer. Reconsider the costs / risk factors of the job elements to bring pricing closer to the ideal time and materials.

Site Visit—an in-person visit by the estimator to the location of the work to be done. The estimator will evaluate access, take measurements, determine components to be retained or removed and scope of work. Estimators will also evaluate risk factors and buyer gullibility and use them to set pricing.

Swindler – From the start he has no intention of doing the agreed-upon job in a competent manner for the price negotiated. He intends to mess with one or more of those. He may: abscond with the down payment; work to the first milestone and then lose steam; use junky materials that don't hold up; employ inexperienced workers who do sucky first-timer work; perform teardown and then demand more money to continue; slap on extra charges left and right; invent costly problems. Even if he ends up completing the work, the total price may exceed the highest quote you received, but without that guy's quality level.

Truck fee – shorthand for usual and customary expenses for a business where the workers come to your establishment in a truck or van. Truck fee might include travel time, equipment rental, perishable tools and supplies, disposable safety equipment, warranty, and so on. Truck fee + labor + materials + risk + permits = customer price.

Verbals—Shorthand for spoken promises or assurances. May be anything the estimator says, one of the employees says, a subcontractor says, or something the hiring party requested and received an affirmative. Verbals should be noted in the contract, either in the body or in an Addendum. Any verbals that are not written there are not part of the deal, and are unenforceable and uncompensated if not done.

Warranty—usually for one year, sometimes longer, from completion date. Make a callback to set an appointment or secure replacement.

Addendum A –Forms

Daniel Homeowner
123 Wonderful Lane
Cherry Hill MA, 01800
(414) 555-1234
DH123@xxx.com

Request for Quote for Sidewalk Replacement

February 14, 2015

Please email your response by:
February 25th, 2015
To <u>DH123@xxx.com</u>
Note: work may start after April 15th and take no longer than 2 weeks start to finish. Must be complete before June 19th.

(contractor)
(contractor address)
(contractor email address/phone)

Dear Sir or Madam:

Please provide a quote to remove the existing sidewalk from street to the front door and replace it with colored and stamped concrete. If interested in quoting call after 6 PM to set up an appointment to view the terrain and show us the types of finishes, colors and looks we can choose from. Tentatively we are thinking of a Lannon stone look with brown highlights.

Scope: The existing sidewalk is 80 ft. long and 3 feet wide; we would like to go 4 feet wide. Also, the original sidewalk bent around a tree but that tree is gone now so we would like a more direct path. We would prefer a natural curve around the corner of the garage rather than the current angle. The existing slab at the front door is to be removed and a one-step high porch that is 5 ft. x 7 ft. in size, with finish and color matching the sidewalk is desired. By use of gentle slopes less than 6" per ten feet we want the current 7 steps to be reduced to 4 steps that are only 5" high. One of the steps would be at the porch and the other three beside the garage. Homeowner will provide and install the handrail affixed to the garage.

Please provide a preliminary quote with the following 7 items:

1) Cost per linear foot of 4" minimum thick, colored through and through, stamped concrete with rebar or wire mesh as needed using 4000 psi concrete with 6% air-entrainment conforming to ASTM C 260 with a consistent slump between 4 to 5 inch. Do not use admixtures containing calcium chloride. Use superplasticizer to maintain consistent slump. Includes subgrade compacting, gravel, and dampening. After stamping, applying appropriate and sufficient water-retaining curing compound. Includes returning 30-60 days later in accordance with mfr's instructions for removal of release agent (if any)

and application of StoneSaver or equivalent low to medium-gloss sealer mixed with ground up polyethylene anti-slip (SharkGrip™) applied with rollers. Cost per linear foot includes compacted gravel base 4" minimum, however depth should be appropriate for prevailing ground conditions.

2) Cost per step with 5" rise and 18-20" tread.

3) Cost for tearing out and disposing of the existing sidewalk and porch, including 6" minimum deep excavation of gravel & compacted earth where the new sidewalk will not follow the same path as the old one.

4) Cost for filling these gaps plus building up the lawn around the sides of the new sidewalk with good loam and hydroseeding with drought-tolerant, slow growing grass mixture.

5) Addresses of three locations where you have done similar colored/stamped concrete sidewalks/patios. Prefer one of them to be over 5 years old.

6) Terms of your warranty. We will ask to see proof of Comprehensive General Liability insurance and Worker's Compensation Insurance coverage prior to start.

7) Estimated start and finish date. We require the pour date to be within two weeks of the start of removal of the old sidewalk and the project completely finished except for the application of StoneSaver and Sharkgrip by June 19th.

Preference is for work to take place and be completed in May.

Additional information:

We require you to contact dig-safe prior to starting work. We have drawings showing the path of sewer and plumbing lines, however we cannot vouch for their accuracy so count on you to take all due precautions.

One person on the crew should be fluent in English.

Your workers may use the downstairs restroom off the kitchen during the project, about 10 ft. from the back door. We will lay a tarp and request they stay on the tarp.

A list of material vendors for this project is to be provided prior to start.

The contractor is to furnish all materials, labor, tools, equipment, and supervision necessary to complete the work described in this specification. A waiver of lien form from the major suppliers is required prior to the holdback payment.

Shield shrubs and perennials near the house from damage; stakes may be used to prop them away as needed. Advise before start of work if one or more need to be moved temporarily.

Cable lines and power line to the gas light: Where these go under the sidewalk the wire is to be inserted inside a 2" diameter PVC pipe. I will contact the cable company and electrician and cover the cost of those service calls, if any. Notify me immediately upon finding a wire and have PVC piping available for laying in prior to their arrival.

All work shall be performed in accordance with all applicable laws, codes, ordinances, and regulations of all local, state, and federal government agencies. It will be the responsibility of the contractor to purchase or obtain all necessary certificates, permits, and licenses required by such agencies, and to give us copies.

Damage to the driveway, curbs, house or garage to be repaired at contractor's expense.

We would like a contract with 10% due at signing and 60% upon completion of sidewalk. Another 20% will be payable after loam and hydroseeding is complete, and the final 10% payable after receipt of lien releases and within 5 days after the removal of release agent per manufacturer's instructions and application of Sharkgrip per manufacturer's instructions.

Regards,

Daniel Homeowner

Daniel Homeowner
123 Wonderful Lane
Cherry Hill MA, 01800
(414) 555-1234
DH123@xxx.com

June 19, 2015

Please email or mail your response by:
June 28, 2015
To DH123@xxx.com or 414-555-1234

(contractor)
(contractor address)
(contractor email address)

Request for Quote for Window Replacement

Dear Sir or Madam:

Please provide a quote for replacing 6 windows at my house. The current windows are Anderson windows and the new ones are to match, with the exception noted below. Three of them face East and three face South. Please phone to set up a site visit. Contractor will be responsible for all measurements and fit.

Contractor to supply all labor and equipment to remove and legally dispose of the existing windows and install the replacement in accordance with the manufacturer's instructions. Materials, tools and disposal are to be included in the price.

Please plan to start work on a Monday and be finished by Friday of the same week. No weekend work is allowed. Work to be performed in July or August. Include the start date in your quote. Provide a 60 hour notice if the start date changes after contract signing.

If contractor encounters rot damage or other issue requiring additional work he is to notify homeowner for review of the site prior to proceeding. Please include the hourly rate at which additional work will be performed.

Specifications:
Remove 5 existing windows and replace with new Anderson A series casement windows white outside and oak inside that fit the openings. Hardware to be brushed chrome. Install ice and water shield around windows and fill any gaps between the existing framing and new window with wood shims prior to replacing the trim.

Painting is not required; if any painting is needed the homeowner will do it later. Contractor to repair interior drywall damage or gaps and endeavor to keep exterior undamaged. I have a few pieces of siding should it be needed. I also have one 8 foot ladder for outside and one 3-step stool for inside use.

One window will have the opening enlarged and an arched-top A series window installed per manufacturer's instructions. Exterior color to be white with oak inside and brushed chrome hardware.

Contractor to furnish proof of liability insurance and workman's comp insurance at signing of contract.

Payment schedule to be $100 down at signing and payment in full within 10 days of receipt of lien release from window supplier after completion.

Regards,

Daniel Homeowner

Right of Rescission Form

Your Right to Cancel

You Could Lose Your Home	[Identification of the type of transaction giving rise to the right of rescission]
	[A statement of the security interest and risk to home]
Your Right to Cancel	[A statement of the right to cancel the transaction giving rise to the right of rescission]
	[As applicable, a statement about the creditor's delay in making funds available to the consumer.]
If You Cancel	[Statements about the creditor not charging a cancellation fee and about refunding fees.]
	[As applicable, statements about the effect of the cancellation on the existing line of credit]
How to Cancel	[Statements about how a consumer may exercise the right to cancel.]
Deadline to Cancel	[A statement of the calendar date on which the three-business-day period for rescission expires.]
	[A statement that the right to cancel the transaction may extend beyond the stated date and a statement of how a consumer may exercise the right to cancel in that case]

[At the creditor's option, a statement that joint homeowners may have the right to rescind and that a rescission by one owner is effective for all owners.]

[At the creditor's option, a format for the consumer to use to acknowledge receipt of the notice, as follows:

Initial here _____ *to acknowledge the receipt of this notice on* _____ .]
  [initials]   *[date]*

[A form for exercise of the consumer's right to cancel, as follows:]
cut here → -

I AM CANCELLING THIS [*identify type of transaction giving rise to the right of rescission*].

_____
Name

_____
Property Address
[12345678_____]
[Account Number]

Addendum A – Forms

Several Samples of Lien Releases

General Principles

Claimant: Supplier, Contractor or Subcontractor

No lien release is binding unless the actual claimant executes (signs) the form.

Paying your primary contractor or getting a release signed by him does not guarantee that other claimants, like subcontractors and suppliers, are paid.

A claimant is a person who, if not paid, can take you to court to get their money and file a lien on your home.

The waiver and release forms can be one of the following:
- Conditional Waiver And Release Upon Progress Payment
- Unconditional Waiver And Release Upon Progress Payment
- Conditional Waiver And Release Upon Final Payment
- Unconditional Waiver And Release Upon Final Payment

The conditional ones are useless to you as a homeowner; they are only for business-to-business deals. Only the unconditional ones protect you.

These forms are binding if signed by the claimant or his or her authorized agent. Be certain the signer is a person with authority to represent their employer in this fashion. One of the workers is often not that person.

This website has lien releases for all states that you can print out and give to your GC or contractor at the start of the project, since he usually professes to not have any copies or have never seen one before:

http://www.zlien.com/lien-waivers/lien-waiver-forms/

The forms for some states are skimpy or incomplete. It is OK to hunt for a better form and use that one. The strength lies in the words. Use a form that says what you want it to say, or even write it yourself, and it will be perfect and legally binding.

Progress payment Release. When the GC must pay suppliers/subs for their work up to this point, whether they are done or not. One signed by each sub and supplier to vouch, 'Yep, I was paid $X amount by the GC.'

UNCONDITIONAL WAIVER AND RELEASE ON PROGRESS PAYMENT

NOTICE TO CLAIMANT: THIS DOCUMENT WAIVES AND RELEASES LIEN, STOP PAYMENT NOTICE, AND PAYMENT BOND RIGHTS UNCONDITIONALLY AND STATES THAT YOU HAVE BEEN PAID FOR GIVING UP THOSE RIGHTS. THIS DOCUMENT IS ENFORCEABLE AGAINST YOU IF YOU SIGN IT, EVEN IF YOU HAVE NOT BEEN PAID. IF YOU HAVE NOT BEEN PAID, USE A CONDITIONAL WAIVER AND RELEASE FORM.

Identifying Information

Name of Claimant:	
Name of Customer:	
Job Location:	
Owner:	
Through Date:	

Unconditional Waiver and Release

This document waives and releases lien, stop payment notice, and payment bond rights the claimant has for labor and service provided, and equipment and material delivered, to the customer on this job through the Through Date of this document. Rights based upon labor or service provided, or equipment or material delivered, pursuant to a written change order that has been fully executed by the parties prior to the date that this document is signed by the claimant, are waived and released by this document, unless listed as an Exception below. The claimant has received the following progress payment:

$_____

Exceptions

This document does not affect any of the following:
(1) Retentions.
(2) Extras for which the claimant has not received payment.
(3) Contract rights, including (A) a right based on rescission, abandonment, or breach of contract, and (B) the right to recover compensation for work not compensated by the payment.

Signature

Claimant's Signature:	
Claimant's Title:	
Date of Signature:	

California lien release

UNCONDITIONAL WAIVER AND RELEASE ON FINAL PAYMENT

NOTICE TO CLAIMANT: THIS DOCUMENT WAIVES AND RELEASES LIEN, STOP PAYMENT NOTICE, AND PAYMENT BOND RIGHTS UNCONDITIONALLY AND STATES THAT YOU HAVE BEEN PAID FOR GIVING UP THOSE RIGHTS. THIS DOCUMENT IS ENFORCEABLE AGAINST YOU IF YOU SIGN IT, EVEN IF YOU HAVE NOT BEEN PAID. IF YOU HAVE NOT BEEN PAID, USE A CONDITIONAL WAIVER AND RELEASE FORM.

Identifying Information

Name of Claimant:	
Name of Customer:	
Job Location:	
Owner:	

Unconditional Waiver and Release

This document waives and releases lien, stop payment notice, and payment bond rights the claimant has for all labor and service provided, and equipment and material delivered, to the customer on this job. Rights based upon labor or service provided, or equipment or material delivered, pursuant to a written change order that has been fully executed by the parties prior to the date that this document is signed by the claimant, are waived and released by this document, unless listed as an Exception below. The claimant has been paid in full.

Exceptions

This document does not affect any of the following:
Disputed claims for extras in the amount of: $

Signature

Claimant's Signature:	
Claimant's Title:	
Date of Signature:	

One used by several states:

CONTRACTOR'S FINAL RELEASE AND WAIVER OF LIEN

Project/ Owner	Contractor
Project: _____	Name: _____
Address: _____	Address: _____
Owner: _____	Contractor Licence: _____
	Contract Date: ___ / ___ / ___

TO ALL WHOM IT MAY CONCERN:

For good and valuable consideration, the receipt and sufficiency of which is hereby acknowledged, the undersigned Contractor hereby waives, discharges, and releases any and all liens, claims, and rights to liens against the above-mentioned project, and any and all other property owned by or the title to which is in the name of the above-referenced Owner and against any and all funds of the Owner appropriated or available for the
construction of said project, and any and all warrants drawn upon or issued against any such funds or monies, which the undersigned Contractor may have or may hereafter acquire or possess as a result of the furnishing of labor, materials, and/or equipment, and the performance of Work by the Contractor on or in connection with said project, whether under and pursuant to the above-mentioned contract between the Contractor and the Owner pertaining to said project or otherwise, and which said liens, claims or rights of lien may arise and exist.

The undersigned further hereby acknowledges that the sum of _____
Dollars ($_____) constitutes the entire **unpaid** balance due the undersigned in connection with said project whether under said contract or otherwise and that the payment of said sum to the Contractor will constitute payment in full and will fully satisfy any and all liens, claims, and demands which the Contractor may have or assert against the Owner in connection with said contract or project.

Dated this ___ day of _____ 20___

Witness to Signature: Contractor

_____   By: _____

_____   Title: _____

Washington state lien release
(if you want to use this one, cross out the 'Washington' and write your own state in ink.)

WAIVER OF LIEN BY CONTRACTOR, SUBCONTRACTOR(S) AND SUPPLIER

We, the undersigned, acknowledge receipt of the amounts stated below as full payment for all labor, professional services, materials, or equipment furnished for use on or about the property of _____ (owner) in _____ County, Washington, through the _____ day of _____ (month), 20__ (year). The property is described as follows (give legal description):

_____
_____
_____
_____

Each person or entity signing this release form releases and waives any interest in the property described above and releases and waives any right to claim a lien on that property for any labor, professional services, materials, or equipment provided through the date listed above. Each person or entity signing this release form reserves the right to claim a lien for any labor, professional services, materials, or equipment provided after that date, to the extent allowed by law.

The consideration received by each person or entity for this release is as follows:

COMPANY NAME	AUTHORIZED SIGNATURE	$ AMOUNT RECEIVED	
PRINTED NAME OF PERSON SIGNING THIS RELEASE	TITLE	DATE	☐ CONTRACTOR ☐ SUBCONTRACTOR ☐ SUPPLIER
COMPANY NAME	AUTHORIZED SIGNATURE	$ AMOUNT RECEIVED	
PRINTED NAME OF PERSON SIGNING THIS RELEASE	TITLE	DATE	☐ CONTRACTOR ☐ SUBCONTRACTOR ☐ SUPPLIER
COMPANY NAME	AUTHORIZED SIGNATURE	$ AMOUNT RECEIVED	
PRINTED NAME OF PERSON SIGNING THIS RELEASE	TITLE	DATE	☐ CONTRACTOR ☐ SUBCONTRACTOR ☐ SUPPLIER
COMPANY NAME	AUTHORIZED SIGNATURE	$ AMOUNT RECEIVED	
PRINTED NAME OF PERSON SIGNING THIS RELEASE	TITLE	DATE	☐ CONTRACTOR ☐ SUBCONTRACTOR ☐ SUPPLIER

This Lien Release form is provided by Labor & Industries as required under RCW 60.04.250.

Michigan state lien release

Michigan folks, this is exactly what is meant by some of the forms are better than others. You might want to go hunting for a better form in me/one of the other states.

There isn't even a line for the dollar amount or work duration! Be sure it is written on the page, because an unconditional waiver for no stated amount and no stated date range doesn't carry any water.

FULL UNCONDITIONAL WAIVER

My/our contract with _____ to provide _____ for the improvement of the property described as: _____
_____
_____
having been fully paid and satisfied, all my/our construction lien rights against such property are hereby waived and released.

If the improvement is provided to property that is a residential structure and if the owner or lessee of the property or the owner's or lessee's designee has received a notice of furnishing from me/one of us or if I/we are not required to provide one, and the owner, lessee, or designee has not received this waiver directly from me/one of us, the owner, lessee, or designee may not rely upon it without contacting me/one of us, either in writing, by telephone, or personally, to verify that it is authentic.

_____
(signature of lien claimant)

Signed on: _____ Address: _____
_____
Telephone: _____

DO NOT SIGN BLANK OR INCOMPLETE FORMS. RETAIN A COPY.

Addendum B – Risk

When the estimator visits, whether he's conscious of it or not he is assessing risk He includes risk either as a percentage of the total job cost or a dollar amount. This amount included in the initial quote means his price will not rise at every complication or additional cost.

While the project is unpredictable to you, that's not true of the skilled tradesman. Most of the time he has foreseen the usual gang of complications and put money into the contract for those.

Do not ask him how much extra it will cost when he brings a snag to your attention. While he may report the difficulty to you because it changes the sequence or completion time, every little thing does not increase the price.

Play for your own team. His job is to monitor his expenses to prevent cost overrun, and your job is to give a frowny face and discourage all cost overruns. Make him specifically ask, in writing, via amendment to the contract, for the extra funds, regardless of the nature of the snag.

If you can reduce any of these risk factors, tell the estimator during the initial visit. Make sure he clearly knows how it will be addressed, along with anything else you are going to do yourself to lower the cost. Cost and risk reductions that he doesn't know about will not lower your price.

The Ten Risk Factors

Customer risk: litigious; poor; miscommunication; indecision

Logistics risk: transporting new product to the customer, removing old; oversize loads, physical impediments to smooth workflow

Installation risk: running into existing problems that stop work or increase scope beyond the skill set of the hired firm; infra-structure not as expected; issues with fit into available space

Product risk: missing parts; incorrect for application; restocking fees; wrong color or 'look' entailing doing it twice; odds of backorders/out of stock pushing deadlines

Legal risk: objections from neighbors; local and state ordinances; property lines; doing tasks requiring licensing or certifications

Environment risk: weather; insufficient light, excessive heat or cold; days when work cannot take place

Adjacent risk: chances of damaging surroundings or items unrelated to project; loss or damage to tools

Design risk: unproven designer; odds that some element of the design doesn't work or meet the needs

Safety risk: protective equipment usage (often disposables); hazards that slow down work; fumes; odds of injury; poor footing/impediments in the work area; visibility issues; pets

End of job risk: cleanup costs; removal of old materials; disposal costs; segregation of hazardous material

Customer risk

Poor is a risk factor. "Pleading poverty" may work with relatives, but not with people who depend upon you to pay them.

Regardless of the work, hinting that you are short of cash is a way to make them lose interest, not gain a better price. This is not a sophisticated insight; a kid won't even rake your leaves if you tell him you have no money. The problem with making sure everyone hears about a lack of funds is that every bidder might highball, so all there is to pick from are outrageously high quotes.

A business has less recourse against a non-paying customer than you might think. Taking the non-paying customer to court doesn't mean the customer now magically has the money they didn't have before. So the reasonable way to avoid losses is to not take the job, before materials and labor costs are incurred.

Mortgage companies make a study of ability to pay, and the guy who needs the lowest payment is charged the highest interest rates. Even when it's plain old money, the richer you are the less you pay. The more conscientious you are, the less you pay. Conscientious and financially secure earns the best deals. Not just at the bank, everywhere. With everybody.

Therefore it is especially important for customers living in a poor neighborhood or a changing neighborhood to give the estimator the impression they're good for it and will have no trouble paying the bill. Guard against going overboard talking about money, which will look preoccupied with affording it.

Living in a less-well-off neighborhood does not mean you're doomed to overpay. There's poor and then there's living beneath one's means. People living beneath their means are intriguing, and when discovered are admired and respected.

Regardless of the neighborhood, it's a safe estimate that one out of ten households are living beneath their means. Eventually they may move to another neighborhood within their means, but for now they live here and need some work done. This is a demographic that can also get a good price.

Businesses want solvent customers. They can't ask nosy questions. They decide based on what you wear, what you say, and if they come to your home or business, how well you maintain your stuff.

When I was six years old, my mom heard that an airline gave free flights to the Lennon Sisters, a popular singing family on the Lawrence Welk show, back in the days when a round trip cross-country ticket cost the equivalent of two week's pay for a skilled tradesman. Mom was peeved that famous, well-paid people would get something for free that they could certainly afford. My sister, seven, pondered this for a while, then grew excited. "Mom! Let's pretend we're rich and famous so they will give us free flights too!" I remember it only because my mother retold the story so many times.

Being rich and famous isn't necessary to get a price break. A little bit more well off, a little bit more connected, and the price is lower.

Years ago I moved from a small contract shop to a Fortune 500 company. I arrived with my Rolodex of good vendors and contract shops I had done business with for years. My purchasing world changed. When working at the contract shop if I asked for three pieces I'd be charged the single part price. Now the exact same sales rep eagerly offered to send the 'samples' for free.

I was honest. I said "Really, there's no bigger order behind this one, I only need three, just let me know the price." It made no difference; they were tickled pink to have a contact at the huge corporation. Sometimes when the box of free samples was opened, inside I'd find a coffee mug and a pack of post-it notes to boot! I didn't feel like a rich and famous company, just little ol' me who bought from this vendor a dozen times before. Overnight, apparently my last name changed to Lennon.

In the business world there's an impression that wealthier people and firms are smarter, have rich friends interested in the same work, are less inclined to go to small claims court for tiny dollar amounts and most of all, are more likely to be repeat customers and pay their bills on time. What's not to like?

Another type of customer risk is miscommunication and indecision. This isn't about being hesitant and using the wrong word in a sentence or not knowing the right term.

Don't say 'yes', or 'fine' or the affirmative 'uh-huh', or even nod when you are not in agreement or when you don't understand. Do not play devil's advocate or be vague on your preferences. Avoid thinking out loud when you have not yet made a decision. These all mislead the contractor, which will cost you big money in restocking fees and tearouts.

If you find yourself nodding while the contractor is talking, be sure to follow it with clear and unambiguous words: "Thanks for all this information, let me think about it and get back to you." Recap or summarize your understanding of the conversation so there is no inadvertent impression of approval for courses of action to which you do not feel committed.

The last but not least customer risk is being litigious. There is simply no reason to talk about taking him to court while still at the estimating stage. Making threats is not the way to begin a working relationship. Anyone who advocates talking about suing someone right in the beginning as a way to put them on a short leash is rude; guys may quote the job in spite of that behavior, against their better judgment, but don't kid yourself that anything is improved by indicating how litigious one is.

Logistics risk:

Logistics refers to transporting and moving things to the worksite and away, as well as moving materials around the worksite. Material handling, trucking, carrying or overcoming obstacles which add cost or more physical labor are part of logistics risk. The estimator must determine what additional time and equipment above the normal range are needed, and calculate that cost. For instance, if usually a small truck or motorized vehicle is used but due to terrain it will have to be done with wheelbarrows or hand dollies here, additional labor cost will result.

Before the estimator arrives, think about access and share your thoughts without being asked or requiring the estimator to bring it up. For example, if an estimator can't spot a parking space for the van, he will add extra labor to cart materials from a block away without even mentioning it. If you bring up that you have a great relationship with the next door neighbor who will allow the van to be parked on his front lawn for this project, that charge goes away.

Look at things through the estimator's eyes. If you see any logistics matters that might slow him down or require extra equipment, mention it even if you think he might already know it.

Sometimes doing teardown or removal yourself prior to the morning of the install can defray costs. Don't however, remove the old one without mentioning it and getting the OK for that, regardless of what 'the old one' is. This is because seeing how the old one is oriented or connected up can sometimes make the installation quicker. There could be a part or two from the existing setup they need to retain or not damage during removal. If you want to remove the old one to save some money, always have a discussion longer than one sentence on that with the estimator.

Engage him to respond, affirm there will be a price break. If you merely say it aloud, he might miss the meaning of your words.

Installation risk

This risk applies when something new is being installed into something existing. If there are doubts about whether it will fit or if modifications can be made that will enable it to work, the estimator will add fudge factors into the quote. This amount is there to cover some trial and error in the installation. In theory if the installation goes well there should be a refund of the labor that wasn't expended, but most contractors just pocket it. Their view is that sometimes it takes longer and sometimes shorter, it averages out for them.

If you think the estimator might be adding installation fudge factors, ask the estimator to speak about the variables possible with this job. Sometimes a very attentive listener who repeats back what he heard can roll the conversation around to asking for the base charge to be for the cheapest choice with an immediate OK for added cost if one of the other variables crops up, which will be paid at itemized time and materials.

Running into problems that either stop work, require extra supplies and tools, or increase the scope of their work beyond the skill set of the workers on site can add cost to the contractor. There's a certain amount of labor, perhaps a few hours, that most contractors will eat without asking for a price change. But if it looks like the odds are good for larger amounts of time, say it's an old house or the previous owner did a lot of DIY projects, they'll put several extra hours in the quote so they don't have to ding the customer for minor cost overruns.

A customer can make a huge difference in the size of the quote if they can address installation risk. Perhaps you can allow a hole to be poked or other measure taken so a certain condition can be ruled out, or share facts about the history or the nature of 'what's back there.' Sometimes the quote assumes protecting something or taking more care than you require. During the quote process for my new fireplace I emphasized they forget protecting the rug or cutting it neatly, because the rug was being replaced right after the fireplace was done. "Clean your putty knives off on it if you like" was how I made the point.

Product risk

This consists of missing parts on equipment, finding the wrong model or type was ordered, or the right one ordered but the wrong one delivered. Each of these can bring work to a halt. It could entail restocking fees if the product that will work couldn't be determined ahead of time. Perhaps the product has the wrong color or look than was expected. Also, the risk that an out of stock condition at the distributor or store could delay other work on the project. The cost due to time waste might be added in the middle of the job.

Legal risk

This risk doesn't include being sued by the customer; that is under customer risk. The legal risk mentioned here pertains to local and state ordinances, respecting property lines, following code, and ensuring the employees doing the work have the proper licenses. It also entails potential extra time in obtaining the proper approvals, permits and certifications for the job.

Another type of legal risk relates to the odds of being sued by neighbors or abutters during the course of the work or after. If some neighbor doesn't understand the rights or where the edges of the licensing requirements are, they might just call the building inspector to make hassle where none needed to be.

There is often a fixed cost associated with notifying abutters (people owning property touching the property being worked on), pulling permits, getting inspections, and so on. If an estimator spots a lot of do-it-yourself work already in place that perhaps goes beyond what is usually allowed, so will the inspector and this might hang up the project—although legally it will be way more trouble for the homeowner than the contractor, it lengthens the time to the end of the job, meaning his next customer's job starts late.

When it comes to legal matters, an ounce of prevention is worth a pound of cure. Tell neighbors ahead of time, read up on permit requirements, and ask questions before work starts.

Environment risk

Weather, insufficient light, excessive heat or cold creates days when work cannot take place at the expected pace. An estimator may use a percent, such as 5% of total job cost, for slowdowns due to heat in summer, cold in winter, rain, and other conditions.

Addressing working conditions by bringing in heaters, adding light, adding cover, providing cold water, and other measures can help the workers work optimally and reduce labor. Mentioning what you intend to do to mitigate environment risk at the estimating stage is valuable.

Adjacent risk

Adjacent risk is when there is inadvertent damage to the surroundings or the object being repaired; it also can apply to the cost of replacing tools that were lost or damaged during the job. Expensive damage will be covered by insurance (you checked they had it, right?) but minor damage will just be fixed or replaced.

If the workers damage something, bite back the urge to be critical. Showing your frowny face is OK.

Unfortunately, most people get a head of steam and immediately threaten small claims court because that feels so good at the moment. Keep that big gun holstered until much later. A contractor may need a third party estimate of the damage in some cases before he can even discuss a resolution. In others, the repair cost may be determined on the spot.

In some cases it works best if the customer proposes what a fair reimbursement amount would be after doing the legwork on getting ballpark figures from third parties, rather than leave it to him to pull something out of the air.

It reduces time and trouble for both parties if the homeowner can say I feel $$ is a fair amount, or I can get it repaired or replaced for that, whatever applies. This method prevents hard feelings if the loss is something without an obvious price tag. For instance, say the situation is that the workers didn't warn the homeowner of the hazard their fumes might pose to small animals so now the ferret is dead.

There are those who advocate throwing out a higher number at first because the guy will counteroffer. My view is people get greedy, develop an opportunist glow in their eyes and ask for some huge amount for their grief and pain that the contractor MUST now dispute. We've all been around the block and know greed when we see it. But don't be a push-over if his proposal isn't close to the real cost to remedy.

On adjacent damage the only fair and equitable resolution is making the victim whole, meaning full reimbursement restoring the customer to their previous state regarding that area or item.

In the case of the ferret, the homeowner may have spent $1,500 on medical treatment for it last year but that doesn't mean the ferret is worth $2,000. Double the replacement cost with a minimum of $200 might be reasonable, and the contractor might even dock some of it from the involved worker's pay, helping them to remember to warn customers to remove all their pets in the future.

Provide a well-thought out price, having two quotes to show if possible, and even offering details regarding the repair required if that is what will happen. That way he can see if he counter-offers with a lesser amount the customer will be out of pocket, and also that any small claims court would settle for the requested amount. He cuts his losses and most likely agrees to the asked-for amount.

Here's a thought: consider reimbursement in other-than-cash form. Perhaps an upgrade, some extra labor on the job, or doing a small something on your mom's property. You get the idea, be creative and truly think about what this contractor could do for you or people you care about that will have a much higher value to you than cost to him. This makes the best of a bad situation for both of you. A homeowner could get that ceramic-tiled bathroom floor she always wanted, and on reflection she didn't want a new ferret.

Another course of action is to deduct the repairs or replacement cost from the amount not yet paid, and pay only the remainder. Consider this a land mine that should not be triggered until the job is complete. If you find his first or second thoughts not quite where you want them to be, tabling this matter puts completing the job front and center.

What if the job ends and the conversation doesn't take place, or you can't seem to get someone on the phone to talk about it? Dock the final payment amount, but remember to include a note with the final payment that you are deducting for repairs, and include a copy (not original) of the invoice or bill related to doing the repairs. The numbers should match to the penny. Usually the cooler heads in Accounting will prevail. If you don't write a clear letter and just underpay, now Accounting is riled too.

There are cases where the contractor's employee successfully hides the damage and you don't discover it until after the job is done. As mentioned above, get your ducks in a row regarding your desired resolution, then call the company. Whether you want to start out talking to the owner, or the accountant, or the person who did the damage is your choice based upon the knowledge you have at that point. One of the three is the optimal course, but which one may not be clear to you.

Design risk:

This is the risk a shop or contractor perceives when dealing with an unproven architect or designer, or a design that seems poorly done based on the skilled tradesman's experience and professional judgment. Poor design causes a lot of rework and mid-stream changes—and that's if you're lucky and it's caught when the build is still in progress because the alternative is it isn't caught and becomes a catastrophic failure later. Either way the cost is paid by the customer, but the contractor incurs more time demand which can delay and add cost to other projects.

Be wary about siding with a Designer or Architect against the contractor in knee-jerk fashion. When a contractor gives a little push-back on some aspect they feel is difficult to build, or a layout that doesn't work with the surroundings or a feature that needlessly adds cost for no end-product value, it is ill-advised for a customer to reply "Make it like the print." A contractor's objections always deserve serious consideration.

Don't assume the guy with the most years is always right. Really look at the parties involved and really think about which side deserves the final say on this particular issue.

Safety risk

Depending upon what is being done, safety risk can be minimal or significant. For instance, building a house has plenty of risks, but building one on a steep incline overlooking a lake has more risk. Access problems, poor footing/uneven ground, fumes, visibility issues, more exposure to sharp edges, dogs on the premises, and a hundred other things can mean a greater expense in personal protective equipment (PPE) and may slow down work. A single injury that causes work stoppage that day and perhaps a few lost days of work might move the job out of profitability for the contractor.

An estimator can't add the full impact of a potential (but not actual, yet) accident into the cost, but most estimators do add a risk factor multiplier to the job. In some cases of known hazards, employers give a hazard pay premium to those working on the job. That pay bump is passed to the customer.

Efforts to improve safety should be communicated to the estimator in the very beginning. Ask him what you can do to improve worker safety; it can lower the price.

End of job risk:

This pertains to after the job is done, presumably satisfactorily, and now the tidying up and disposal actions come into play. This is the part the average contractor or installer begrudges, since he places his value on his skills in doing the job. This part often entails skilled people doing unskilled labor, however, it is the last thing the customer sees. It carries a larger proportion of the customer's feel for job quality than it should, somewhat like the dismount from apparatus in gymnastics competitions. A poor dismount weighs overly large in the scoring, and a fall on that final tumbling pass can trick memory into thinking the whole thing was unskilled.

I have seen chimney sweeps and house painters thrive and prosper with advertising that said little about their skill set, but mainly asserted they were the best at cleaning up when they were done. Sold!

There's a lot to be said for advertising good end-of-job performance and then coming through; guys who have the self-discipline to do the last part well are probably also doing the rest of it well. Character will out. We're playing the odds of a good outcome, and having no other information to go on, the tidy fellow is going to get more points for high quality work than the sloppy fellow. He nails the landing and that counts for a lot.

End of job risk, beyond cleanup, also includes disposal of old parts and waste materials. Sometimes these materials must be separated and go to different disposal or recycling sites. Oils will go to one place, hazardous waste to another, dirt removed to another, wood to another, and so on. This is a lot of trucking and handling and the cost is forwarded to the customer. It's a cost of doing business and all of it is covered by the price the customer pays. There's no escaping it.

You might try saving some money by offering to have them leave without cleaning on completion instead of sweeping and wiping down, but don't be offended if they refuse. They know the lingering impression it leaves and can't risk their reputation.

Index

Addendum to the contract, 77
Adjacent risk, 222
air time, 73
amending the contract, 107
Angie's List, 131
architect, 147
attic, 161
bank loans, 97
bathrooms, 166
Better Business Bureau, 131
Big Dogs vs. Small Fry, 89
Black Walnut tree, 186
boilerplate, 102
bonding, 112
boring
    as motivation, 184
boxers and bar fights, 27
bread-and-butter, 4, 11, 56
Building Inspector, 126
Building Inspector, aka BI, 126
callback, 99, 125, 130, 197
cash payments, 37
change notices, 145
check for $200
    unmet expectations, 128
confidentiality,
    implied, on a quote, 90
contract performance bond, 112
Contractor heaven, 34
cost charts, 186
credit card, 135
credit card court, 136

dealbreakers, 79
debit card, 134, 198
Design risk, 223
Do It Yourself
    DIY, 2
doorknobs, 161
Drywall, 177
easy work, 56
egress window
    basement, 164
End of job risk, 224
Environment risk, 221
estimator, 50
Ethics, 87
expectations
    unmet, cause for lawsuits, 128
fines, within a contract, 107
freezing pipes,
    preventing, 164
future business, 57
General Contractors, 142
good contract, purpose of, 101
handwriting
    the law loves, 108
handyman, list of jobs, 48
high points
    contract including only, 101
highballing, 32
HOA
    Homeowner's Association, 141
holdback, 125
idle time, 73, 74

Installation risk, 220
jargon, 54
job duration, 66
keeper of the verbals, 105
lawsuits, causes of, 128
Leadtime, 138
Lennon Sisters, 218
Level scheduling, 55
license bond, 112
licensing
    contractor, 111
lien release, 113
Logistics, 219
milestones, 96
Money Magazine, 1
mulligan, 133, 199
Negotiation skills, 9
non-disparagement clause, 104
opportunism
    preventing, 26
    typical motivation, 25
    when strongest, 38
Opportunism
    single quote, 76
OSB, 181
outlets, 156
overtime, 66
per-foot averages, 68
personality
    as a tool, 185
pets, during construction, 119
plaster, 177
poor
    as a risk factor, 217
Product risk, 221
professional organizations, 45
project cost, reducing, 146
punch list, 125

repercussions, contractual, 106
Request for Quote- RFQ, 76
Rescission, Right of, 106
reverse estimating, 70
RFQ
    followup, 81
    obedience, 82
rich and famous, 218
risk
    the ten, 216
roof overhang, 178
roofs, 177
Safety risk, 224
Schadenfreude, 184
scope, reducing, 146
sealed bid
    federal government, 91
second quote phenomenon, 37
sewage-ejector systems
    for toilet, 164
signoff at completion, 99
small claims court, 128
steps and stairways, 171
Stronghold Crusader, 193
swindler, 31
time and materials, 60, 68, 220
tweaking in
    after completion, 130
unconditional waiver
    lien release, 113
unit pricing, 69
up-flushing toilet, 164
verbal promise, 105
verbals
    keeper of the, 105
warranty work, 136
Will, must, shall, 110
zoning laws, 141

Some other books by this author:

Getting the Best Price on a Used Car

http://www.amazon.com/dp/B00CE2DSVA

Amazon's top-selling car-buying book. How to hunt for a car while not playing the dealership's game.

Flying for First Timers

http://www.amazon.com/dp/B00I9P0VO8

Tips to make airline travel less of a hassle, whether you haven't flown in years or fly only twice a year. Sections on traveling with pets, special needs traveling and packing.

Italy for First Timers

http://www.amazon.com/dp/B00I9P0VO8

Up to date information on travel to Italy--from booking the flight to coping with lost luggage, and everything between. Restaurants, hotels, tours, trains, VAT tax and more. Pros and cons of package tours. Tons of tips to make this trip enjoyable and safe.

Warranty Work: Getting a Good Outcome Over the Phone

http://www.amazon.com/dp/B00MX8P26O

A step by step guide for accomplishing the warranty call in the proper spirit to make a good outcome almost inevitable.

Briefcase Travel: Packing Light for Business Trips

http://www.amazon.com/dp/B00N84BWV6

Travel with only a briefcase or large purse on your next two or three day trip–without buying anything new!

Index

www.ingramcontent.com/pod-product-compliance
Lightning Source LLC
Chambersburg PA
CBHW020754160426
43192CB00006B/326